Magnetohydrodynamics:
Waves and Shock Waves in Curved Space-Time

MATHEMATICAL PHYSICS STUDIES

Series Editor:

VOLUME 14

Magnetohydrodynamics: Waves and Shock Waves in Curved Space-Time

by

André Lichnerowicz

Chaire de Physique Mathématique,
Collège de France,
Paris, France

KLUWER ACADEMIC PUBLISHERS
DORDRECHT / BOSTON / LONDON

Library of Congress Cataloging-in-Publication Data

Lichnerowicz, André, 1915-
Magnetohydrodynamics : waves and shock waves in curved space-time / by André Lichnerowicz.
p. cm. -- (Mathematical physics studies ; v. 14)

1. Magnetohydrodynamics. 2. Shock waves. 3. Calculus of tensors. 4. Mathematical physics. I. Title. II. Series.
QC718.5.M36L53 1994
538'.6--dc20 94-7970
CIP

ISBN 978-90-481-4390-0

Published by Kluwer Academic Publishers,
P.O. Box 17, 3300 AA Dordrecht, The Netherlands.

Kluwer Academic Publishers incorporates
the publishing programmes of
D. Reidel, Martinus Nijhoff, Dr W. Junk and MTP Press.

Sold and distributed in the U.S.A. and Canada
by Kluwer Academic Publishers,
101 Philip Drive, Norwell, MA 02061, U.S.A.

In all other countries, sold and distributed
by Kluwer Academic Publishers Group,
P.O. Box 322, 3300 AH Dordrecht, The Netherlands.

Printed on acid-free paper

Table of Contents

Chapter III - Einstein's equations and the Leray theorem.

Chapter IV - Gravitational and electromagnetic shock waves.

Chapter V - Relativistic hydrodynamics.

Chapter VI - The equations of magnetohydrodynamics.

Chapter VII - Magnetohydrodynamic shock waves.

Chapter VIII - Hugoniot's function and applications.

Annex I - Shock waves and alfven waves.

Annex II - Magnetosonic rays.

Annex III - Classical approximations of the relativistic shock equations.

Note - Approach to a quantum theory of fields for a curved space-time.

II - Applications to quantization problems over a curved space-time.

PREFACE

For seventy years, we have known that Einstein's theory is essentially a theory of propagation of waves for the gravitational field. Confusion enters, however, through the fact that the word ***wave,*** in physics, implies sometimes repetition and sometimes not. This confusion is often increased by he use of Fourier transforms, by which a disturbanse which appears to be without repetition is resolved into periodic wave-trains with all frequencies. But, in a general curved space-time, we have nothing corresponding to Fourier transforms.

Here, we consider systematically waves corresponding to the propagation of discontinuities of physical quantities describing either fields (essentially electromagnetic fields and gravitational field), or the motion of a fluid, or together, in magnetohydrodynamics, the changes in time of a field and of a fluid.

The main equations, for the different studied phenomena, constitute a ***hyperbolic system*** and the study of a formal Cauchy problem is possible. We call ***ordinary waves*** the case in which the derivative of superior order appearing in the system are discontinuous at the traverse of a hypersurface, the wave front ; we call shock waves the case where the derivatives of an order inferior by one are discontinuous at the traverse of a wave front.

From 1950, many well-known scientits (Taub, Synge, Choquet-Bruhat, etc.) have studied the corresponding equations for different physical phenomena : systems associated to the electromagnetic and gravitational fields, to hydrodynamics and to magnetohydrodynamics.

The purpose of this book -which is a book of mathematical physics- is to give a synthesis of the results and to precise some new results, using a unique method, the method of ***the tensor-distributions.*** In particular, this method is a powerful tool for the study of the gravitational shock waves and for the magnetohydrodynamical shock waves, where a complete theory may be given. Such a theory is specially adapted to astrophysics, in particular for the treatment of the pulsars.

The same method may be used for the quantization of some fields in a curved space-time. Therefore, we have included in the book a Note concerning an approach of the quantization of fields in this context.

We hope that this book is clear and contains sufficiently detailed mathematical proofs, with the suitable mathematical rigor. One of its purposes is to give an easy access to the different works on this subject during half-century and also to most recent results.

A. Lichnerowicz
June 1993

I - Tensor-distributions.

The purpose of Chapter I is to introduce to the mathematical formulation of tensor distributions and to show how this formulation can furnish a simple and precise tool for the study of waves in mathematical physics, and particularly in general relativity. The author introduced this formalism systematically in 1960 for study of the quantum theory of fields over a curved space-time. We shall apply here this tool to problems on waves relative to Maxwell's equations and to Einstein's equations and, in the same way, to waves that arise in relativistic hydrodynamics and magnetohydrodynamics.

I.1.- Tensor distributions over a Pseudo-Riemannian manifold.

a) Let V_{n+1} be a connected, paracompact, ***oriented,*** differentiable manifold of dimension $n+1$ and class of differentiation $C^{h+1} (h \geq 1)$. We define over V_{n+1} a pseudo-riemannian "metric" given by a non degenerate 2-tensor g of class C^k $(0 \leq k \leq h)$ and of arbitrary signature. In a local chart $\{x^\alpha\}$ $(\alpha, \beta, \ldots = 0, 1, \ldots, n)$ of V_{n+1} with domain Ω, we may write

$$g|_\Omega = g_{\alpha\beta}\, dx^\alpha \otimes dx^\beta$$

If T and U are two p-tensors at the point x of V_{n+1}, we denote by $(T, U)(x)$ the scalar product at x of T and U. In the domain Ω, one has

(I.1.1) $$(T, U)(x) = (T_{\alpha_1 \ldots \alpha_p} U^{\alpha_1 \ldots \alpha_p})(x) \qquad (x \in \Omega)$$

Let $\mathcal{D}^p(V_{n+1})$ be the space of p-tensors U of V_{n+1}, of class C^k with compact supports S(U). If T is a locally summable p-tensor over V_{n+1}, we may put for all $U \in \mathcal{D}^p(V_{n+1})$

$$(I.1.2) \qquad < T , U > = \int_{V_{n+1}} (T, U)(x)\, \eta(x)$$

where η is the natural volume element of the manifold. Locally

$$(I.1.3) \qquad \eta|_\Omega = \sqrt{|g|}\, dx^0 \wedge dx^1 \wedge \ldots \wedge dx^n$$

A p-tensor distribution T ***of*** V_{n+1} ***is a continuous linear form, with scalar values, over the space*** $\mathcal{D}^p(V_{n+1})$

If $U \in \mathcal{D}^p(V_{n+1})$, T [U] or < T, U > represents the value of T for the test tensor U. In the preceding definition, ***continuous*** is to be understood in the usual sense of the theory of distributions. Given a compact set K, one considers the sequence $\{U_\nu\}$ ($\nu = 1, 2, \ldots$) of elements of $\mathcal{D}^p(V_{n+1})$ with supports $S(U_\nu) \subset K$, such that the U_ν and all their derivatives of order $\leqslant h$ converge uniformly to 0 as $\nu \to \infty$, T is ***continuous*** if $T[U_\nu]$ tends to 0 for all these sequences, the supports belonging to an arbitrary compact K. For antisymmetric tensors, one has an identification with currents in the sense of G de Rham.

b) If u is a scalar distribution and V a p-tensor of class C^h denote by u . V the p-tensor distribution defined by

$$(u . V)[U] = u[(V , U)] \qquad U \in \mathcal{D}^p(V_{n+1})$$

Now let Ω be the domain of a local chart $\{ x^\alpha \}$. Given, over Ω, n^p scalar distributions $T_{\alpha_1 \dots \alpha_p}$, one may define a p-tensor distribution over Ω

(I.1.4) $$T_{\alpha_1 \dots \alpha_p} \, dx^{\alpha_1} \otimes \dots \otimes dx^{\alpha_p}$$

in the following way : if $\{ e_\alpha \}$ is the frame with dual coframe $\{ dx^\alpha \}$, consider an element U of $\mathcal{D}^p(\Omega)$ given by

$$U = U^{\alpha_1 \dots \alpha_p} \, e_{\alpha_1} \otimes \dots \otimes e_{\alpha_p}$$

$T[U]$ is given by the sum :

$$T[U] = T_{\alpha_1 \dots \alpha_p} \, [\, U^{\alpha_1 \dots \alpha_p} \,]$$

with summation with respect to $(\alpha_1 \dots \alpha_p)$

Conversely, any p-tensor-distribution T over Ω may be described by such an expression . Let $T_{\alpha_1 \dots \alpha_p}$ be the scalar distribution defined by

$$T_{\alpha_1 \dots \alpha_p} [\, f \,] = T [\, f \, e_{\alpha_1} \otimes \dots \otimes e_{\alpha_p} \,] \qquad f \in \mathcal{D}^p(\Omega)$$

One has for $U \in \mathcal{D}^p(\Omega)$

$$T[U] = T[\, U^{\alpha_1 \dots \alpha_p} \, e_{\alpha_1} \otimes \dots \otimes e_{\alpha_p} \,] = T_{\alpha_1 \dots \alpha_p} [\, U^{\alpha_1 \dots \alpha_p} \,]$$

and T can be described by (I.1.4)

Thus, if T is a p-tensor distribution of V_{n+1}, the restriction of T to the domain Ω of a local chart admits, in this chart, components that are scalar-distributions over Ω.

I.2. - Covariant derivative of a Tensor Distribution.

We suppose for the moment $h, k \geqslant 1$

Let ∇ be the covariant derivation operator defined by the pseudo-riemannian connection determined by $\mathbf{g}$. We denote by ε the codifferentiation operator over tensors : if U is a $(p+1)$-tensor, εU is the p-tensor given, in a local chart, by

$$\varepsilon : U = \{ U^{\alpha_0 \cdots \alpha_p} \} \rightarrow \varepsilon U = \{ - \nabla_\alpha U^{\alpha\alpha_1 \cdots \alpha_p} \}$$

a) Let T be a p-tensor of $\mathbf{V}_{n+1}$ admitting a locally summable covariant derivative. If Ω is the domain of a chart $\{ x^\alpha \}$ and if $U \in \mathcal{D}^{p+1}(\Omega)$, one has :

$$< \nabla T \,, U > = \int_{V_{n+1}} \nabla_\alpha T_{\alpha,\dots\alpha_p} U^{\alpha\alpha_1 - \alpha_p}$$

or, by integration by parts

$$< \nabla T, U > = \int_{V_{n+1}} \nabla_{\alpha_0} (T_{\alpha_1 \dots \alpha_p} U^{\alpha_0 \alpha_1 \dots \alpha_p}) - \int_{V_{n+1}} T_{\alpha_1 \dots \alpha_p} \nabla_{\alpha_0} U^{\alpha_0 \alpha_1 \dots \alpha_p}$$

The first term on the right-hand is the integral of the divergence of a vector with compact support and is zero according to the Stokes formula. it follows that

$$< \nabla T , U > = < T , \varepsilon U > \qquad (U \in \mathcal{D}^{p+1} (\Omega))$$

Now let $U \in \mathcal{D}^{p+1} (\nabla_{n+1})$. Introduce a finite coverning $\{ \Omega_\nu \}$ of an open neighborhood of $S(U)$ by domains of local charts and let $\{ \varphi_\nu \}$ be a subordinate partition of unity. Set $U_\nu = \varphi_\nu U$ so that $U = \Sigma U_\nu$, $\varepsilon U = \Sigma \varepsilon U_\nu$. One has

$$< \nabla T , U_\nu > = < T , \varepsilon U_\nu >$$

and by summation over ν

(I.2.1) $$< \nabla T , U > = < T , \varepsilon U > \qquad (U \in \mathcal{D}^{p+1} (V_{n+i}))$$

We are thus led to define the covariant derivative of a p-tensor distribution T ***as the*** $(p+1)$***-tensor distribution*** ∇T ***determined by relation*** (I.2.1.)

b) If u is a scalar distribution, the preceding definition gives $\nabla u = du$, where du is the usual differential of scalar distribution. Let T be a vector distribution (1-tensor distribution). Over the domain of a chart (I.2.1) may be written

(I.2.2) $$< \nabla_\alpha T_\beta , U^{\alpha\beta} > = - < T_\beta , \nabla_\alpha U^{\alpha\beta} > \qquad (U \in \mathcal{D}^2 (\Omega))$$

Let $\{ \Gamma^\alpha_{\beta\gamma} \}$ be the coefficients of the connection in our chart. One has (with $\partial_\alpha = \partial / \partial x^\alpha$)

$$\nabla_\alpha U^{\alpha\beta} = \partial_\alpha U^{\alpha\beta} + \Gamma^\alpha_{\varrho\alpha} U^{\alpha\beta} + \Gamma^\beta_{\varrho\alpha} U^{\alpha\varrho}$$

and as a consequence :

$$< \nabla_\alpha T_\beta , U^{\alpha\beta} > = - < T_\beta , \partial_\alpha U^{\alpha\beta} + \Gamma^\alpha_{\varrho\alpha} U^{\varrho\beta} > - < T_\beta , 1^\beta_{\varrho\alpha} U^{\varrho\beta} >$$

If we fix the index β , it follows from the definition of $\partial_\alpha T_\beta$ that

$$- < T_\beta \, , \partial_\alpha U^{\alpha\beta} + \Gamma^\alpha_{\rho\alpha} U^{\rho\beta} > = < \partial_\alpha T_\beta \, , U^{\alpha\beta} >$$

and we obtain :

$$< \nabla_\alpha T_\beta \, , U^{\alpha\beta} > = < \partial_\alpha T_\beta - \Gamma^\rho_{\beta\alpha} T_\rho \, ; U^{\alpha\beta} >$$

We have established

(I.2.3) $$\nabla_\alpha T_\beta = \partial_\alpha T_\beta - \Gamma^\rho_{\beta\alpha} T_\rho$$

More generally, it may be established that the classical properties of the covariant derivation in a pseudo riemannian connection and all the corresponding formulas are valid for tensor distributions. In particular, if T is a $(p+1)$-tensor distribution and U an ordinary p-tensor such that $S(T) \cap S(U)$ is compact, one has

(I.2.4) $$< \varepsilon T \, , U > = < T \, , \nabla U >$$

I.3. - The distributions Y^+ , Y^- , and $\overline{\delta}$ corresponding to a hypersurface.

Let Σ be a regular hypersurface of $\mathbf{V}_{n+1}$. It may be defined locally by the equation $\varphi = 0$ (where φ is supposed to be of class C^2) ; say that Σ is regular, that is that $d\varphi(x) \neq 0$ for any $x \in \Sigma$. For brevity, we put $\ell = d\varphi$.

a) We restrict ourselves to a contractile domain of $\mathbf{V}_{n+1}$, that is partionned by Σ into two domains Ω^+ and Ω^- corresponding respectively to $\varphi > 0$ and $\varphi < 0$. Let Y^+_φ , or more briefly Y^+ (resp. Y^-_φ or Y^-) be a function over Ω that is equal to 1 (resp. 0) over Ω^+

and to 0 (resp. 1) over Ω^- . These functions define over Ω scalar distributions, again denoted Y^+ , Y^- with respect to the formulas

(I.3.1) $$< Y^+ , f > = \int_\Omega Y^+ f \eta = \int_{\Omega^+} f \eta \quad (f \in \mathcal{D}^0 (\Omega))$$

and

(I.3.2) $$< Y^- , f > = \int_\Omega Y^- f \eta = \int_{\Omega^+} f \eta \quad (f \in \mathcal{D}^0 (\Omega))$$

Consider the class of Leray n-forms associated with φ , that is the n-forms satisfying the relation :

(I.3.3) $$\eta = d\varphi \wedge \alpha = \ell \wedge \alpha$$

If α , α' are two forms of this class, there exists an (n-1)-form β such that

(I.3.4) $$\alpha' = \alpha + d\varphi \wedge \beta$$

We denote by $\partial\Omega^+$ and $\partial\Omega^-$ the oriented boundaries over Σ of Ω^+ and Ω^- (with $\partial\Omega^+ = - \partial\Omega^-$). According to (I.3.4) the integral

$$\int_{\partial\Omega^-} f\alpha = - \int_{\partial\Omega^+} f\alpha \qquad (f \in \mathcal{D}^0 (\Omega))$$

has a well-determined value independant of the choice of α satisfying (I.3.3). Thus one may define a scalar distribution $\overline{\delta}_\varphi$ or $\overline{\delta}$, the Dirac distribution over Ω by the relation

(I.3.5) $$< \overline{\delta} , f > = \int_{\partial\Omega^-} f\alpha = - \int_{\partial\Omega^+} f\alpha \qquad (f \in \mathcal{D}^0 (\Omega))$$

$\overline{\delta}$ has its support over Σ .

b) It is often convenient to introduce over Ω a local chart $\{ y^\alpha \}$ said to be adapted to Σ, that is such that $y^0 = \varphi$. In such a chart, ℓ admits covariant components $\ell_0 = 1$, $\ell_i = 0$ (i any Latin index $= 1, \dots, n$).

We seek the vector distributions that are the derivatives of the scalar distributions Y^+ and Y^-. In an adapted chart

(I.3.6) $$\eta|_\Omega = \sqrt{|g|}\, dy^0 \wedge dy^1 \wedge \dots \wedge dy^n$$

It follows that the n-form of Ω

(I.3.7) $$\alpha = \forall|g|\; dy^1 \wedge \dots \wedge dy^n$$

satisfies (I.3.3). If $U \in \mathcal{D}^1(\Omega)$, one has according to the definition of covariant derivation

$$< \nabla y^+ , U > = < Y^+ , \delta U >$$

In an adapted chart $\{ y^\alpha \}$ with domain Ω, it follows that

$$< \nabla Y^+, U > = < Y^+, \frac{1}{\sqrt{|g|}} \partial_\alpha (U^\alpha \sqrt{|g|}) =$$

$$= - \int_{\Omega^+} \partial_\alpha (U^\alpha \sqrt{|g|})\, dy^0 \wedge dy^2 \wedge \dots \wedge dy^n$$

From Stocke's formula, the last integral is equal to

$$- \int_{\partial\Omega^+} U^0 \sqrt{|g|}\, dy^1 \wedge \dots \wedge dy^n = - \int_{\partial\Omega^+} \ell_\alpha U^\alpha \alpha$$

From (I.3.5) one obtains

$$< \nabla Y^{+}, U > = < \ell\, \overline{\delta}, U > \qquad (U \in \mathscr{D}^{1}(\Omega))$$

Then in the sense of distributions

(I.3.8) $$\nabla Y^{+} = \ell\, \overline{\delta}$$

and similarly

(I.3.9) $$\nabla Y^{-} = - \ell\, \overline{\delta}$$

c) We study the vector distribution that is the differential of $\overline{\delta}$. If V is a vector of Ω , one has

$$< i(V)\, \overline{\delta}, f > = < \nabla\, \overline{\delta}, f V > = < \overline{\delta}, \varepsilon(f V) > \qquad f \in \mathscr{D}^{0}(\Omega)$$

suppose that V is tangent to hypersurfaces φ = const. $(i (V) \ell = 0)$. In adapted coordinates, we have $V^{0} = 0$ and

$$< i(V)\, \nabla \overline{\delta}, f > = \int_{\partial\Omega+} \partial_{\alpha}(f V^{\alpha} \sqrt{(|g|} \ ) \ dy' \wedge ... \wedge dy^{n}$$

let

$$< i(V)\, \nabla\, \overline{\delta}, f > \ \int_{\partial\Omega+} \partial_{i} (f V^{i} \sqrt{|g|} \ ) \ dy' \wedge ... \wedge dy^{n}$$

From Stocke's formula, the last integral is equal to

$$\int_{\Omega} + \partial_{i}\, \partial_{0} (f V^{i} \sqrt{(|g|}) \ dy^{0} \wedge dy' \wedge ... \wedge dy^{n}$$

f having compact support, one sees by integration with respect to y^{i} of the corresponding term that each term of this integral is zero. Thus, for $i(V)\ \ell = 0$, one has $i(V)\ \nabla\ \overline{\delta} = 0$, that is in adapted coordinates $\nabla_{i}\, \overline{\delta} = 0$. If we put $\nabla_{0}\, \overline{\delta} = \overline{\delta}\,'$, we see that there exists over Ω a scalar distribution $\overline{\delta}\,'$ such that

(I.3.10) $$\nabla\overline{\delta} = \ell \, . \, \overline{\delta}'$$

$\overline{\delta}'$ has its support over Σ

I.4.- Tensor discontinuities relative to a hypersurface.

a) Let T be a p-tensor over Ω that is of class C^0 over Ω^+ and Ω^-. One says that T is regularly discontinuous at the traverse of Σ when, if φ tends to 0 through positive (resp. negative) values, T converges uniformly to a function with tensor values over Σ denoted T^+ (resp T^-). The discontinuity of T is the tensor defined over Σ given by $[T] = T^+ - T^-$. The tensor distributions Y^+T and Y^-T are defined over Ω in a natural manner.

T is a locally summable tensor ; if T^D is the tensor distribution defined by T, we have in the sense of distributions

(I.4.1) $$T^D = Y^+ T + Y^- T$$

Similarly the tensor distributions $\overline{\delta}\,T^+$, $\overline{\delta}\,T^-$ may be defined with supports over Σ and

$$\overline{\delta}\,[\,T\,] = \overline{\delta}\,T^+ - \overline{\delta}\,T^-$$

b) Now, for Ω, let T be a p-tensor of class C^1 over Ω^+ and Ω^- *such that* T *and* ∇T *are regularly discontinuous at the traverse of* Σ ; ∇T defines a tensor-distributions $(\nabla T)^D$ given by

(I.4.2) $$(\nabla T)^D = Y^+ \nabla T + Y^- \nabla T$$

We seek to evaluate the tensor ∇T^D, the covariant derivative of the tenor-distribution T^D given by (I.4.1). We have

$$\nabla T^D = \nabla(Y^+T) + \nabla(Y^- T)$$

From our definitions and (I.3.8), it is easy to show that

(I.4.3) $$\nabla(Y^+T) = \ell\, \overline{\delta} \otimes T^+ + Y^+ \nabla T$$

Similarly

(I.4.4) $$\nabla(Y^- T) = -\ell\, \overline{\delta} \otimes T^- + Y^- \nabla T$$

By addition, we obtain acording to (I.4.2) :

(I.4.5) $$\nabla T^D = \ell \otimes \overline{\delta}\,[\,T\,] + (\nabla T)^D$$

In this formula ∇T^D is the tensor distribution that is the covariant derivative, in the sense of distributions, of the tensor-distribution T^D. The tensor $(\nabla T)^D$ is the tensor distribution defined by the usual covariant derivative ∇T of the tensor T. The term $\ell \otimes \overline{\delta}\,[T]$ is often called the tensor 0-layer corresponding to the tensor discontinuity $[T]$ over Σ

c) We study the covariant derivative of the tensor distribution $\overline{\delta}\,[T]$. It may be verified immediately that in adapted coordinates

$$\nabla_i(\bar{\delta}\,[T]\,) = (\nabla_i\,\bar{\delta}\,) \otimes [T] + \bar{\delta}\,\nabla_i\,[T] = \bar{\delta}\,\nabla_i\,[T]$$

But from the uniform convergence $\nabla_i\,[T] = [\,\nabla_i\,T\,]$. One sees that there exists a p-tensor distribution δT with support over Σ such that

(I.4.6) $$\bar{\delta}\,[\,\nabla T\,] = \nabla\,(\bar{\delta}\,[T]\,) + \ell \otimes \delta T$$

In particular for a scalar f satisfying the above hypotheses over the tensor T

(I.4.7) $$\bar{\delta}\,[\,df\,] = d\,(\bar{\delta}\,[\,f\,]\,) + \ell\,\delta f$$

I.5. - Generalization of ∇T^D

The definition of ∇T^D and formula (I.4.5) may be extended to more general circonstances and this generalization is useful for the theory of gravitational shock waves.

We suppose for a moment that $k = 1$ over Ω^+ and Ω^-, but not over Ω. More precisely we suppose that the first derivatives of the components of the metric tensor g are regularly discontinuous at the traverse of Σ, that is that $\mathbf{g}$ is piecewise C^1 relatively to Σ

a) Let

$$\omega = (\,\omega^\alpha_\beta\,) = (\,\Gamma^\alpha_{\beta\gamma}\,dx^\gamma\,)$$

the 1-form of connection, where the Γ are the coefficients of the connection in the chart $\{x^\alpha\}$ of domain Ω. It is clear that ω is continuous over Ω^+ and Ω^- and regularly discontinuous as the traverse of Σ.

Let T be a vector of class C^1 over Ω^+ and Ω^- such that T and ∇T are ***regularly discontinuous*** at the traverse of Σ. We set by definition

$$\nabla T^D = dT^D + Y^+(\omega T) + Y^-(\omega T) \tag{I.5.1}$$

It is easy to see that one thus defines a tensor distribution over Ω and that, if ω is continuous at the traverse of Σ, this definition of ∇T^D coincides with the preceding definition (Section II.2)). Formula (I.5.1) may be extended in a natural manner to tensors of arbitrary order, and the operator ∇ thus admits the usual properties of the covariant derivation operator

b) We proceed to establish.

Proposition. -*Suppose that* g *is piecewise* C^1 *relative to* Σ. *If* T *is a tensor of class* C^1 *over* Ω^+ *and* Ω^-, *such that* T *and* ∇T *are regularly discontinuous, formula* (I.4.5) *is again satisfied*

To lighten the notations, we present the proof for the case where T is a vector. Choose in particular chart with domain Ω. One may apply (I.4.5) to scalars defined over Ω by the various components of T. There results

$$dT^D = \ell \otimes \bar{\delta}[T] + (dT)^D$$

and it follows from (I.5.1)

$$\nabla T^D = \ell \otimes \overline{\delta}\,[T] + (dT)^D + Y^+\,(\omega T) + Y^-\,(\omega T)$$

But over each domain Ω^+ or Ω^-

$$\nabla T = dT + \omega T$$

and ∇T defines over Ω the tensor distribution

$$(\nabla T)^D = (dT)^D + (\omega T)^D = (dT)^D + Y^+(\omega T) + Y^-(\omega T)$$

One obtains

(I.5.2) $$\nabla T^D = \ell \otimes \overline{\delta}\,M + (\nabla T)^D$$

that is formula (I.4.5) again

I.6. - The case of continuous tensors.

We return to the hypothesis that the tensor g ***is everywhere of class*** C^k $(k \geqslant 1)$

a) We now consider tensors T that are ***continuous over*** Ω ***, of class*** C^1 ***over*** Ω^+ ***and*** Ω^-***, such that*** ∇T ***is regularly discontinuous*** at the traverse of Σ . These tensors define in a natural manner a tensor algebra A for the tensor product. Formula (I.4.6) may be writen

(I.6.1) $$\overline{\delta}\,[\nabla T] = \ell \otimes \delta T$$

consider the map $\delta : T \in A \to \delta T$, where δT is a tensor distribution with support over Σ . From (I.6.1), δ is a derivation of the tensor algebra $A : \delta$ is linear and if $T , U \in A$, it follows from (I.6.1) that

$$\delta(T \otimes U) = \delta T \otimes U + T \otimes \delta U$$

δT is called the infinitesimal discontinuity of T. According to its defintion, δT depends only on T and not on the connection.

b) In the remainder of this section, we suppose $h \geqslant 2$, $k \geqslant 1$ and the metric is assumed ***piecewise*** C^2 at the traverse of Σ.

Let $\Gamma^{\alpha}_{\beta\gamma}$ be the coefficients of the connection Γ with respect to a chart $\{ x^{\alpha} \}$ with domain Ω. Under the hypothesis $k \geqslant 1$, these coefficients are continuous and define local scalars on Ω. According to (I.4.7), there exist scalar distributions $\delta\Gamma^{\alpha}_{\beta\gamma}$ over Ω such that

(I.6.2) $$\overline{\delta}\,[\,\partial_{\rho}\,\Gamma^{\alpha}_{\beta\gamma}\,] = \ell_{\rho}\,\delta\,\Gamma^{\alpha}_{\beta\gamma}$$

But h being $\geqslant 2$, the $[\partial_{\rho}\,\Gamma^{\alpha}_{\beta\gamma}]$ define a tensor over Σ, with respect to changes of charts. Therefore the $\delta\Gamma^{\alpha}_{\beta\gamma}$ are the components of ***a tenor-distribution*** $\delta\Gamma$ over Ω with support over Σ.

Consider the curvature tensor of the connection. According to (I.6.2) we have

(I.6.3) $$\overline{\delta}\,[\,R^{\lambda}_{\rho,\alpha\beta}\,] = \overline{\delta}\,[\,\partial_{\alpha}\,\Gamma^{\lambda}_{\rho\beta} - \partial_{\beta}\,\Gamma^{\lambda}_{\rho\alpha}\,] = \ell_{\alpha}\,\delta\,\Gamma^{\lambda}_{\rho\beta} - \ell_{\beta}\,\delta\,\Gamma^{\lambda}_{\rho\alpha}$$

c) For the tensors T we suppose

Hypothesis (1.1) : T *is continuous over* Ω *and of class* C^2 *over* Ω^{+} *and* Ω^{-}.

The tensors ∇T *and* $\nabla\nabla T$ *are regularly discontinuous at the traverse of* Σ.

The tensor ∇T itself satisfies the hypothesis corresponding to formula (I.4.6). One has therefore

$$\bar{\delta}[\nabla_\alpha \nabla_\beta T] = \nabla_\alpha (\bar{\delta}[\nabla_\beta T]) + \ell_\alpha \delta \nabla_\beta T$$

and from (I.6.1)

$$\bar{\delta}[\nabla_\alpha \nabla_\beta T] = \nabla_\alpha (\ell_\beta \delta T) + \ell_\alpha \delta \nabla_\beta T$$

There results :

$$\bar{\delta}[\nabla_\alpha \nabla_\beta T] = \nabla_\alpha \ell_\beta \delta T + \ell_\alpha \delta \nabla_\beta T + \ell_\beta \nabla_\alpha \delta T \tag{I.6.4}$$

The Bianchi identity gives

$$[(\nabla_\alpha \nabla_\beta - \nabla_\beta \nabla_\alpha) T^{\lambda_1 \dots \lambda_p}] = \sum_{a=1}^{p} [R_\rho{}^{\lambda_a}{}_{,\alpha\beta}]\, T^{\lambda_1 \dots \rho \dots \lambda_p}$$

It follows from (I.6.3), (I.6.4) that :

$$\ell_\alpha \left(\delta \nabla_\beta T^{\lambda_1 \dots \lambda_p} - \nabla_\beta \delta T^{\lambda_1 \dots \lambda_p} - \sum_{a=1}^{p} \delta \Gamma^{\lambda_a}_{\rho\beta} T^{\lambda_1 \dots \rho \dots \lambda_p} \right)$$

is symmetrical in α and β. Therefore there exists a p-tensor distribution $\bar{T}$ such that

$$\delta \nabla_\beta T^{\lambda_1 \dots \lambda_p} - \nabla_\beta \delta T^{\lambda_1 \dots \lambda_p} - \sum_{a=1}^{p} \delta \Gamma^{\lambda_a}_{\rho\beta} T^{\lambda_1 \dots \rho \dots \lambda_p} = \ell_\beta \bar{T}^{\lambda_1 \dots \lambda_p}$$

$\overline{T}$ has its support over Σ. If we subtitute in (I.6.4), we obtain an interesting formula

$$(I.6.5) \qquad \overline{\delta}[\nabla_\alpha \nabla_\beta T^{\lambda_1 \dots \lambda_p}] = \nabla_\alpha \ell_\beta \delta T^{\lambda_1 \dots \lambda_p} + \ell_\alpha \nabla_\beta \delta T^{\lambda_1 \dots \lambda_p} +$$

$$+ \ell_\beta \nabla_\alpha \delta T^{\lambda_1 \dots \lambda_p} + \ell_\alpha \ell_\beta T^{\lambda_1 \dots \lambda_p} + \ell_\alpha \sum_{a=1}^{p} \delta \Gamma^{\lambda_a}_{\rho\beta} T^{\lambda_1 \dots \rho \dots \lambda_p}$$

If the curvature tensor is continuous at the traverse of Σ, we have

$$(I.6.6) \qquad \overline{\delta}[\nabla_\alpha \nabla_\beta T] = \nabla_\alpha \ell_\beta \delta T + \ell_\alpha \nabla_\beta \delta T + \ell_\beta \nabla_\alpha \delta T + \ell_\alpha \ell_\beta \overline{T}$$

One may deduce from (I.6.6) a useful consequence : choose a fixed chart with domain Ω and adopt over Ω the uniform metric defined by this chart : the coefficients of the connection are then zero. Let f be a scalar satisfying Hypothesis (H) over Ω ; one has in this chart

$$(I.6.7) \qquad \overline{\delta}[\partial_\alpha \partial_\beta f] = \partial_\alpha \ell_\beta \delta f + \ell_\alpha \partial_\beta \delta f + \ell_\beta \partial_\alpha \delta f + \ell_\alpha \ell_\beta \overline{f}$$

II - MAXWELL'S EQUATIONS AND ELECTROMAGNETIC WAVES OVER A CURVED SPACE-TIME.

II.1.- Space-time.

Let (V_4, g) be a space-time, which is here an oriented differentiable manifold of dimension 4, of class $(C^2$, piecewise $C^4)$ provided with a metric tensor g of strictly hyperbolic type (signature + ...) of class $(C^1$, piecewise $C^3)$; η is the volume element of (V_4, g). We suppose thus that in the intersection of the domains of two admissible charts, the coordinates of a point $x \in V_4$ in the first chart are functions of class C^2, with non zero jacobian, of the coordinates of the same point x in the second chart. The third and fourth derivatives of these functions exist, but are only piecewise continuous. We shall see that these hypothesis of differentiability are in strict relation with the field equations of general relativity and, as a consequence, with physics.

One knows that a precision in a C^∞-structure may always be found for such a differentiable structure. But it appears that *such a precision of the* $(C^2$, *piecewise* $C^4)$ *structure of* V_4 *is devoid of any physical significance*

For such a chart $\{x^\alpha\}$ with domain Ω, one has

$$g\,|_\Omega = g_{\alpha\beta}\, dx^\alpha \otimes dx^\beta \quad (\alpha, \beta \text{ any Greek indices} = 0, 1, 2, 3).$$

Similarly, we suppose that the ***gravitational potentials*** $g_{\alpha\beta}$ are functions of class C^1, piecewise C^3).

At any point x of V_4, the vectors X of the tangent space $T_x V_4$ satisfying $g(X,X) = 0$ are called ***isotropes*** and generate a cone C_x of $T_x V_4$, the elementary cone in x. The metric tensor g thus defines a field of cones. The vectors (or directions) interior or over the cone C_x ($g_x(X, X) \geqslant 0$) are called ***temporal,*** the vectors strictly interior to the cone C_x ***strickly temporal ;*** the vectors exterior to the cone C_x ($g_x(X, X) < 0$) are called ***spatial.***

II.2. - Singular 2-form.

a) Consider at a point x of V_4 a 2-form $F \neq 0$ for which there exists a vector $\ell \neq 0$ such that

$$\ell \wedge F = 0 \tag{II.2.1}$$

and

$$i(\ell)\, F = 0 \qquad \text{or} \qquad \ell \wedge * F = 0 \tag{II.2.2}$$

where $i(\ )$ is the inner product and $*$ the adjoint operator over forms. In terms of components, the system (II.2.1), (II.2.2) may be writen

$$S\, \ell_\alpha\, F_{\beta\gamma} = 0 \qquad \ell^\alpha F_{\alpha\beta} = 0 \tag{II.2.3}$$

where S is summation over circular permutations. If this hold, F is said to be ***singular at*** x ***with fundamental vector*** ℓ .

The vector ℓ ***is necessarily isotropic :*** if this were not the case, we would be able to introduce at x an orthonormal frame $\{ \mathbf{V}_{(\alpha)}$ admitting ℓ or a proportional vector as an element ($\ell = \mathbf{V}_{(0)}$; $\mathbf{V}_0^2 = \pm 1$). In this frame, one has from (II.2.3) $F_{ij} = 0$, $F_{0i} = 0$ $(i, j = 1, 2, 3)$ and F would be zero at x . The vector ℓ being chosen, it follows from (II.2.1) that there exists a vector b , defined up to $b \to b + \sigma\ell$ (where σ is a scalar) such that

(II.2.4) $$F = \ell \wedge b$$

Since the vector ℓ is isotropic, one deduces from (II.2.2) that $i(\ell)$ $b = 0$ and b is a spatial vector of the 3-plane π_x tangent to the elementary cone C_x along ℓ . The scalar

(II.2.5) $$e = - b^\alpha b_\alpha = \| b \|^2$$

is ***strictly positive.***

Given F , the direction defined by ℓ is well determined : the vector ℓ is defined up to the change $\ell \to \lambda\ell$ (where λ is a scalar). In fact if ℓ' is another isotropic vector satisfying (II.2.1), one has from (II.2.4) $\ell' : \lambda\ell + \mu b$ and ℓ' is isotropic if and only if $\mu = 0$. We see that the scalar e depends only on F and on the normalisation of ℓ .

b) Let $v_{(1)}$, $v_{(2)}$ be two vectors orthogonal to ℓ such that $v^2_{(1)} = v^2_{(2)} = -1$, $v_{(1)} v_{(2)} = 0$. These vectors define a spatial 2-plane of π_x. One may chose b in this 2-plane so that

$$b = a_1 v_{(1)} + a_2 v_{(2)} = \Sigma \; a_u v_{(u)} \qquad (u = 1, 2)$$

and Equation (II.2.4) can be written

$$F = \sum_u a_u \, (\ell \wedge v_{(u)})$$

or in components

(II.2.6) $$F_{\alpha\beta} = \sum_u a_u (\ell_\alpha v_{(u)\beta} - \ell_\beta v_{(u)\alpha})$$

One has

(II.2.7) $$e = -b^\alpha b_\alpha = \sum_u (a_u)^2$$

The sigular 2-forms play a fundamental role in the representation of electromagnetic waves.

II.3. - Maxwell's equations and the formal Cauchy problem.

a) An electromagnetic field in the absence of induction is described by a skew symmetric tensor or 2-form F (electric fied-magnetic fied), that is in general supposed to be (C^0, piecewise C^2) over V_4. Special relativity lead us to suppose that the electromagnetic field F satisfies over V_4 a tensor system of partial differential equations that is nothing but the translation of Maxwell' equations for a curved space-time. If d is the exterior differentiation operator and ε the codifferentiation, this system can be written

$$dF = 0 \tag{II.3.1}$$

and

$$\varepsilon F = J \tag{II.3.2}$$

where J is the electric current vector, which describes the sources of the electromagnetic field. We suppose in general that j ***is piecewise continuous.*** In domains where $J = 0$ one has what is called the ***exterior case.*** In the following, set

$$D = \varepsilon F \qquad E = \varepsilon (* F) \tag{II.3.3}$$

Maxwell's equations may be written

$$D = J \qquad E = 0 \tag{II.3.4}$$

Since $\varepsilon^2 = 0$ over these forms, D and E satisfy the conservation identities

$$\varepsilon D = 0 \qquad \varepsilon E = 0 \tag{II.3.5}$$

It follows that if J is of class C^1 , Maxwell's equations imply the equation

$$\varepsilon J = 0 \tag{II.3.6}$$

which is called ***the equation of conservation of electricity.***

According to special relativity, one may associate to an electromagnetic field without induction an energy tensor, called ***Maxwell's tensor,*** defined in terms of components by

$$\tau_{\alpha\beta} = \frac{1}{4} g_{\alpha\beta} F_{\lambda\mu} F^{\lambda\mu} - F_{\alpha}^{\varrho} \varrho F_{\beta\varrho} \tag{II.3.7}$$

This tensor has zero trace

$$\tau = g^{\alpha\beta} \tau_{\alpha\beta} = 0$$

We deduce from Maxwell's equations an useful expression for $\nabla_{\alpha} \tau_{\beta}^{\alpha}$. We have

$$\nabla_{\alpha} \tau_{\beta}^{\alpha} = \frac{1}{2} F^{\lambda\mu} \nabla_{\beta} F^{\alpha\varrho} F_{\beta\varrho} - F^{\alpha\varrho} \nabla_{\alpha} F_{\beta\varrho} - \nabla_{\alpha} F^{\alpha\varrho} F_{\beta\varrho}$$

According to (II.3.1), we have

$$F^{\alpha\varrho} \nabla_{\alpha} F_{\beta\varrho} = \frac{1}{2} F^{\alpha\varrho} (\nabla_{\alpha} F_{\beta\varrho} - \nabla_{\varrho} F_{\beta\alpha}) = \frac{1}{2} F^{\alpha\varrho} \cdot \nabla_{\beta} F_{\alpha\varrho}$$

It follows from (II.3.2) that

$$\nabla_{\alpha} \tau_{\beta}^{\alpha} = J^{\varrho} F_{\beta\varrho} \tag{II.3.8}$$

If $J = 0$, Maxwell's tensor is conservative

b) We first mention an elementary analysis of the Cauchy problem for Maxwell's equations in the exterior case. These equations may be written

$$\text{(II.3.9)} \qquad D_\alpha \equiv - g^{\beta\gamma} \nabla_\beta F_{\gamma\alpha} = 0 \qquad\qquad E^\alpha \equiv \frac{1}{2} \eta^{\alpha\beta\gamma\delta} \nabla_\beta F_{\gamma\delta} = 0$$

Let Σ be a local regular hypersurface of V_4 ***not tangent to elementary cones.*** In coordinates $\{ y^\alpha \}$ adaptated to Σ, one has $g^{00} \neq 0$. In such a chart, Maxwell's system is equivalent to the set of two systems

$$\text{(II.3.10)} \qquad \begin{aligned} D_i &\equiv - g^{00} \partial_0 F_{0i} - g^{0k} \partial_0 F_{ki} + G_i = 0 \\ E^i &\equiv \frac{1}{2} \eta^{i0jk} \partial_0 F_{jk} + H^i = 0 \end{aligned}$$

and

$$\text{(II.3.11)} \qquad D^0 = 0 \qquad\qquad E^0 = 0$$

where D^0, E^0, G_i, H^i have known values over Σ if we know the values over Σ of $F_{\alpha\beta}$ (Cauchy data).

Moreover, ***if an electromagnetic field*** F ***satisfies the system*** (II3.10) ***and over only*** Σ, Equations (II.3.11), ***it also satisfies*** (II.3.11) ***outside*** Σ. This is an immediate consequence of the conservation identities (II.3.5).

We see that, for Cauchy data $(F_{\alpha\beta})_\Sigma$ satisfying (II.3.11) over Σ, system (I.3.10) gives the values over Σ of the three derivatives $\partial_0 F_{jk}$ and the three derivatives $\partial_0 F_{0i}$. It follows that, for a solution of Maxwell's equations, the first derivatives of $F_{\alpha\beta}$ may not be discontinuous at the traverse of Σ (since $g^{00} \neq 0$).

We see that the ***characteristic manifolds*** of the Maxwell system under consideration are the hypersurface tangent to the elementary cones. We shall return in this result presently by another method.

II.4. - Maxwell's equations and potential-vector.

a) We introduce the notion of potential-vector. The equation $dF = 0$ expresses that F is locally the exterior differential of a linear form ; over a contractile domain Ω, we have

$$F|_\Omega = d\alpha$$

or in a chart with domain Ω.

$$F_{\lambda\mu} = \partial_\lambda \alpha_\mu - \partial_\mu \alpha_\lambda$$

One says that α ***is a potential vector for*** F ***over*** Ω. Such a potential vector is defined up to a local additive differential. The change $\alpha \to \alpha + dS$ (where S is a scalar over Ω) is called an ***electromagnetic gauge change.*** It is important to note the fact that it is not possible, in general, to define a global potential-vector over all space-time.

To each potential vector α , one associates the scalar $\varepsilon\alpha$. One says that a potential vector α satisfies the ***Lorentz condition*** if $\varepsilon\alpha = 0$. One may always satisfy this condition locally by means of an appropriate gauge change.

b) It is easy to write the Maxwell system in potential vector terms. We have simply over Ω

(II.4.1) $$\varepsilon\, d\alpha = J$$

Introduce the laplacian $\Delta = d\varepsilon + \varepsilon d$ of G . de Rham over the forms. For a 1-form α one has explicitely in terms of coordinates

$$(\Delta\alpha)_\varrho = -\nabla^\lambda \nabla_\lambda \alpha_\varrho + R_\varrho^\sigma\, \alpha_\sigma$$

where R_ϱ^σ is the Ricci tensor of the metric. Equation (II.4.1) may be written

(II.4.2) $$\Delta\alpha - d\varepsilon\alpha = J$$

If α satisfies the Lorentz condition, one has only

(II.4.3) $$\Delta\alpha = J$$

or

(II.4.4) $$-g^{\lambda\mu} \nabla_\lambda \nabla_\mu \alpha_\varrho + R_\varrho^\sigma \alpha_\sigma = J_\varrho$$

c) We propose to study the role of (II.4.3) and of the conservation of electricity equation

(II.4.5) $$\varepsilon J = 0$$

with respect to the Maxwell system. A similar result may be presented for the gravitationnal case.

Let Σ be a spatial hypersurface, $\{ y^\alpha \}$ a chart adapted to Σ ; one has $g^{00} > 0$. In potential vector terms, the Cauchy data over Σ are defined by the components α_λ of α and their first derivatives $\partial_0 \alpha_\lambda$. We have seen that $D^0 = (\varepsilon F)^0$ depends only on the Cauchy data over Σ .

Consider a solution of (II.4.3) (II.4.5) corresponding to given Cauchy data such that

(II.4.6) $$\varepsilon \alpha = 0 \qquad \text{for } y^0 = 0$$

and

(II.4.7) $$(\varepsilon F)^0 = J^0 \qquad \text{for } y^0 = 0$$

First show that one also has

(II.4.8) $$\partial_0 \varepsilon\alpha = 0 \qquad \text{for } y^0 = 0$$

In fact, one has for $y^0 = 0$

$$(d\,\varepsilon\,\alpha)^0 = (\Delta\alpha)^0 - (\varepsilon\, d\,\alpha)^0 = J^0 - (\varepsilon F)^0 = 0$$

But locally over Σ

$$(d\,\varepsilon\,\alpha)^0 = g^{00}\,\partial_0\,\varepsilon\alpha + g^{0i}\,\partial_i\,\varepsilon\alpha = g^{00}\,\partial_0\,\varepsilon\alpha$$

g^{00} being $\neq 0$, one sees that our solution satisfies (II.4.8) over Σ . For this solution, one deduces from (II.4.5) by the action of ε over (II.4.3)

(II.4.9) $$\Delta\,\varepsilon\,\alpha = 0$$

Σ being spatial, the only solution of (II.4.4) satisfying $\varepsilon\alpha = 0$ and $\partial_0\,\varepsilon\alpha = 0$ over Σ is the zero solution. Thus $\varepsilon\alpha = 0$ outside Σ and α satisfies the Lorentz condition. It follows that α is a solution of Maxwell's equation (II.4.1)

Proposition. - *Any solution of* (II.4.3), (II.4.5) *satisfying over the spatial hypersurface* $\Sigma\ (y^0 = 0)$ *the Lorentz condition* $\varepsilon\alpha = 0$ *and the relation* $(\varepsilon F)^0 = J^0$ *is a solution of the corresponding Cauchy problem for the Maxwell equation* (II.4.1) *and the potential vector* α *satisfies the Lorentz condition identically.*

II.5. - Ordinary electromagnetic waves.

We have supposed that the electromagnetic field F is (C^0 , piecewise C^2). We suppose here that the electric current J is continuous.

a) ***An ordinary electromagnetic wave is defined by a solution*** F ***of Maxwell's equations*** (II.3.1) (II.3.2) ***(where*** J ***is continuous) such that the derivatives of*** F ***are regularly discontinuous at the traverse of a hypersurface*** Σ ***which is called the wave front.***

According to (I.6.1), there exists a 2-tensor distribution δF with support over Σ $(\varphi = 0)$ such that

$$\bar{\delta}\,[\nabla F] = \ell \otimes \delta F \qquad (\ell = d\varphi)$$

δF is the infinitesimal discontinuity of F. It is equivalent to say that there exists over Σ a 2-form $H \neq 0$ such that

(II.5.1) $$[\nabla F] = \ell \otimes H$$

where $\delta F = \bar{\delta} \cdot H$. It follows from (II.3.1) that over Σ

(II.5.2) $$\ell \wedge H = 0$$

and from (II.3.2) according to the continuity of J :

(II.5.3) $$i(\ell)\, H = 0$$

The 2-form H is singular, of fundamental vector ℓ , at each point x of Σ and ℓ is necessarily isotropic. We thus have

Proposition. - *The electromagnetic wave fronts (or characteristic manifolds of a Maxwell system) are the hypersurfaces called isotropic tangent at each of their points to the elementary cone at this point.*

b) If Σ is defined by the local equation $\varphi = 0$, the characteristic manifold are solutions of the partial derivative equation of first order, that may be written in an arbitrary chart

(II.5.4) $$\Delta_1 \varphi \equiv g^{\alpha\beta} \partial_\alpha \varphi \, \partial_\beta \varphi = 0$$

ℓ being the differential of φ , one has

(II.5.5) $$\nabla_\alpha \ell_\beta - \nabla_\beta \ell_\alpha = 0$$

ℓ being isotropic :

(II.5.6) $$\ell^\alpha \ell_\alpha = 0$$

and ℓ defines the generator of the elementary cone along which Σ is tangent to the elementary cone. It follows from (II.5.6) that

(II.5.7) $$\ell^\alpha \nabla_\beta \ell_\alpha = 0$$

From (II.5.5) and (II.5.7) one deduces

(II.5.8) $$\ell^\alpha \nabla_\alpha \ell_\beta = 0 \qquad \text{or} \qquad \ell^\alpha \nabla_\alpha \ell^\beta = 0$$

and the trajectories of ℓ are isotropic geodesics of the metric tensor g of space-time, with respect to their affine parameter. These curves are the characteristics of Equation (II.5.4) and are called ***bicharacteristics (or rays)*** of the system of Maxwell's equations.

c) We study the propagation of δF or H along rays. In the interest of simplicity, we suppose that over the domain being considered, J ***is of class*** C^1 ***and*** g ***of class*** C^2 ***, so that the curvature tensor is continuous.*** The tensor ℓ being isotropic, we deduces from (I.6.6) by contraction

(II.5.9) $$\bar{\delta}\,[\nabla^\alpha \nabla_\alpha F] = \pi\,(\ell)\,\delta F$$

where $\pi(\ell)$ is the differential operator

(II.5.10) $$\pi(\ell) = 2\ell^\rho\,\nabla_\rho + \nabla_\rho\,\ell^\rho$$

Over each domain Ω^+ or Ω^- (notation of Section (I.6)) one has the Maxwell equation :

$$\nabla_\alpha F_{\beta\gamma} + \nabla_\beta F_{\gamma\alpha} + \nabla_\gamma F_{\alpha\beta} = 0$$

and by differentiation :

$$\nabla^\alpha \nabla_\alpha F_{\beta\gamma} + \nabla^\alpha \nabla_\beta F_{\gamma\alpha} + \nabla^\alpha \nabla_\gamma F_{\alpha\beta} = 0$$

Subtracting

(II.5.11) $$[\nabla^\alpha \nabla_\alpha F_{\beta\gamma}] + [\nabla^\alpha \nabla_\beta F_{\gamma\alpha}] + [\nabla^\alpha \nabla_\gamma F_{\alpha\beta}] = 0$$

But by virtue of the Ricci identity and the continuity of the curvature tensor, we have

$$[\nabla^{\alpha} \nabla_{\beta} F_{\gamma\alpha}] = [\nabla_{\beta} \nabla^{\alpha} F_{\gamma\alpha}] = [\nabla_{\beta} J_{\gamma}] = 0$$

Similarly

$$[\nabla^{\alpha} \nabla_{\gamma} F_{\alpha\beta}] = [\nabla_{\gamma} \nabla^{\alpha} F_{\alpha\beta}] = - [\nabla_{\gamma} J_{\beta}] = 0$$

There results from (II.5.9)

(II.5.12) $$\pi(\ell)\, \delta F = 2\ell^{\rho} \nabla_{\rho} \delta F + \nabla_{\rho} \ell^{\rho} \delta F = 0$$

According to the property (I.3.10) of $\overline{\delta}$, (II.5.12) is equivalent to

(II.5.13) $$\pi(\ell)\, H_{\alpha\beta} = 2\ell^{\rho} \nabla_{\rho} H_{\alpha\beta} + \nabla_{\rho} \ell^{\rho} H_{\alpha\beta} = 0$$

If H is zero at the point x of Σ, where Σ is tangent to the elementary cone along ℓ, it follows from (II.5.13) that H is zero the entire length of the isotropic geodesic defined by ℓ. One has

Proposition. - *If the electric current* J *is of class* C^1 *and the curvature tensor continuous in a neighborhood of the electromagnetic wave front* Σ, *the infinitesimal discontinuity of the electromagnetic field* F *is propagated along rays that generate* Σ *according to the differential equation*

$$\pi(\ell)\, \delta F_{\alpha\beta} \equiv 2\, \ell^{\rho} \nabla_{\rho} \delta F_{\alpha\beta} + \nabla_{\rho} \ell^{\rho} \delta F_{\alpha\beta} = 0$$

The Proposition expresses the fundamental property of rays.

III. - EINSTEIN'S EQUATIONS AND THE LERAY THEOREM.

III.1. - The system of Einstein equations.

a) Let (V_4, g) be a space-time satisfying the hypothesis of Section (II.1). The first step in the relativistic theory of gravitation is the choice of a tensor system of partial differential equations relating the metric tensor **g** to the energy distributions in space-time.

These energy distributions are the sources of the gravitational field. Einstein was led to these equations by two kinds of considerations : on one hand, these equations must be the extension of the Laplace-Poisson equation that locally relates the newtonian potential and the density of the matter. On the other hand, they have to satisfy a system of four identities, called the conservation identities.

We write these equations in the form

(III.1.1) $$S = \chi T$$

where S and T are two symmetric 2-tensors and χ a physical constant. The tensor T has a purely dynamic significance and describes phenomenologically the momentum-energy distribution at various points of V_4. The ***energy tensor*** T is the extension of the right member of the Poisson equation. It is often assumed to be ***piecewise continuous.*** In a domain where $T = 0$ one has what is termed ***the exterior case.***

The tensor S depends only on the metric structure g of the space-time and satisfies the two conditions.

(1) *The components* $S_{\alpha\beta}$ *of* S *depend only on potentials and on their derivatives of the first two orders ; they are linear with respect to the derivatives of second order.*

(2) *The tensor* S *is conservative, that is satisfy the identity*

(III.1.2) $$\varepsilon S = 0 \qquad \text{or} \qquad \nabla_\alpha S^\alpha{}_\beta = 0$$

To each symmetric 2-tensor h associate over V_4 the symmetric 2-tensor $\mu(h)$ given by

$$\mu(h) = h - (1/2)\,(\mathrm{Tr}\,h\,)\,g$$

The mapping μ is involutive ($\mu^2 = \mathrm{Id}$) and $\mu(h)$ is said to be associated to h.

The Einstein tensor is by definition, the tensor $\mu\,(Ri)$ associated to the Ricci tensor Ri of $(V_4\,,g)$. Elie Cartan has shown that up to a certain additive term λg , the Einstein tensor may be characterized by the two preceding hypothesis. Characterizations by more general hypothesis has been given later. The system of Einstein equations is then the system (III.1.1), where

(III.1.3) $$S = Ri - \frac{1}{2}\,R\,g$$

R is the scalar riemannian curvature

b) The components of the curvature tensor in an arbitrary chart are

$$R^{\alpha}_{\beta,\lambda\mu} = \partial_\lambda \Gamma^{\alpha}_{\beta\mu} - \partial_\mu \Gamma^{\alpha}_{\beta\lambda} + \Gamma^{\alpha}_{\varrho\lambda}\Gamma^{\varrho}_{\beta\mu} - \Gamma^{\alpha}_{\varrho\mu}\Gamma^{\varrho}_{\beta\lambda} \tag{III.1.4}$$

where the coefficients of the pseudoriemannian connection are denoted by $(\Gamma^{\alpha}_{\beta\mu})$. By introducing the Christoffel symbols there results

$$R_{\alpha\beta,\lambda\mu} = \partial_\lambda [\, \beta\mu\, , \alpha\,] - \partial_\mu [\, \beta\lambda, \alpha\,] + K_{\alpha\beta,\lambda\mu}$$

where the $K_{\alpha\beta\ \lambda\mu}$ depend only on potentials and on their first derivatives and are quadratic with respect to these derivatives. By expanding, after simplifications, one obtains

$$R_{\alpha\beta,\lambda\mu} = -\frac{1}{2}\left(\partial_{\alpha\lambda} g_{\beta\mu} + \partial_{\beta\mu} g_{\alpha\lambda} - \partial_{\alpha\mu} g_{\beta\lambda} - \partial_{\beta\lambda} g_{\alpha\mu}\right) + K_{\alpha\beta,\lambda\mu} \tag{III.1.5}$$

The Ricci tensor is defined by

$$R_{\alpha\beta} = R^{\sigma}_{\alpha,\sigma\beta} = g^{\varrho\sigma} R_{\varrho\alpha,\sigma\beta}$$

With (III.1.5) we deduce

$$R_{\alpha\beta} = \frac{1}{2} g^{\varrho\sigma}\left(\partial_{\alpha\varrho} g_{\beta\sigma} + \partial_{\beta\sigma} g_{\alpha\varrho} - \partial_{\alpha\beta} g_{\varrho\sigma} - \partial_{\varrho\sigma} g_{\alpha\beta}\right) + K_{\alpha\beta} \tag{III.1.6}$$

where the $K_{\alpha\beta}$ are quadratic with respect to the first derivatives of the potentials.

c) Determination of the characteristic manifolds of the Einstein system may be carried out by the study of a formal Cauchy problem. Let Σ be a regular hypersurface, $\{ y^\alpha \}$ a chart adapted to Σ. The Cauchy data are the values over Σ on the potentials $g_{\lambda\mu}$ and their first cross partial derivatives $\partial_0 g_{\lambda\mu}$. One has the following results :

(1) The four components S^0_α over Σ of the Einstein tensor depend explicitly only on the Cauchy data and on their tangential derivatives.

(2) Introduce a change of adapted chart according to the formulas

$$y^{\alpha'} = y^\alpha + \frac{1}{2}(y^0)^3 (\varphi^{(\alpha)}(y^i) + \varepsilon^{(\alpha)}(y^\lambda)) \tag{III.1.7}$$

where the $\varepsilon^{(\alpha)}$ converge uniformly to 0 , at the same time as their derivatives, as $y^0 \to 0$. One easily verifies that the coordinate values at each point x of Σ , the values at x of the Cauchy data and the values of the second derivatives $\partial_{00} g_{ij}$ $(i, j = 1, 2, 3)$ are invariant under this change.

But it is possible by choice of $\varphi^{(\alpha)}$ to give arbitrary-values to the four derivatives $\partial_{00} g_{\lambda 0}$. The potentials g_{ij} are called ***significant potentials for*** Σ and the $g_{\lambda 0}$ ***non significant potentials.*** The derivatives $\partial_{00} g_{\lambda 0}$ of non significant potentials are not appear in the expression for the curvature tensor.

According to our general hypothesis on the differentiable structure of V_4 , we may, by choice of $\varphi^{(\alpha)}$ on one hand of Σ and on the other, create or destroy discontinuities that are devoid of physical significance.

3) According to (III.1.6), if $g^{00} \neq 0$ the Einstein system in the exterior case is equivalent to the set of two systems

(III.1.8) $$R_{ij} = -\frac{1}{2} g^{00} \partial_{00} g_{ij} + G_{ij} = 0$$

and

(III.1.9) $$S^0_\alpha = 0$$

where the G_{ij} depend only on the Cauchy data and their tengential derivatives. If a metric satisfies (III.1.8) and, over Σ alone (III.1.9), it necessarily satisfies (III.1.9) outside Σ. This is an immediate consequence of the conservation identities (III.1.2).

4) Suppose that **T** ***is continuous across*** Σ. In order that the second derivatives of significant potentials (or the curvature tensor) be discontinuous at the traverse of Σ, it is necessary that $g^{00} = 0$, that is that Σ be tangent to elementary cones.

An ordinary gravitational wave is defined by a solution g ***to the Einstein equations (where*** T ***is continuous) such that the curvature tensor is regularly discontinuous at the traverse of a hypersurface*** Σ ***, which is called the wave front.***

It is equivalent to say that the second derivatives of the significant potentials for Σ are regularly discontinuous at the traverse of Σ. One may proceed, for these waves, to a study similar to that of Section (II.5.a).

We shall not carry this out, prefering to turn our interest to shock waves, which are more delicate to study and more intersting. We have here

Proposition. *The gravitational wave fronts (or characteristic manifolds of a Einstein system) are isotropic hypersurfaces tangent at each of their points to the elementary cone at that point.*

It follows exactly in the same way as Section (II.5) that bicharacteristics of Einstein systems (or ***gravity rays)*** are again isotropic geodesics of the metric tensor **g** for the space-time. If T is of class C^1, it is easy to prove that one has, along the rays that generate a wave front Σ, the equation of propagation.

$$(\text{III.1.10}) \quad \pi(\ell)\,[\,R_{\alpha\beta,\lambda\mu}\,] \equiv 2\ell^{\rho}\,\nabla_{\rho}\,[\,R_{\alpha\beta,\lambda\mu}\,] + \nabla_{\rho}\,\ell^{\rho}\,[\,R_{\alpha\beta,\lambda\mu}\,] = 0$$

III.2. - Harmonic quantities.

It is convenient for many mathematical studies of Einstein's equations to introduce harmonic charts. We introduce also four quantities, the harmonic quantities, which are the left member of the conditions of harmonicity. We note that harmonic charts are a precious tool for the mathematical analysis of an Einstein system, but afford no special advantage for physical interpretation.

a) Consider a scalar equation of second order admitting the same characteristics as the Einstein system. The simplest equation admitting these properties may be written

(III.2.1) $$\Delta f \equiv \varepsilon\, d f = 0$$

and is the "Laplace equation " for the manifold (V_4, g). Locally

(III.2.2) $$\Delta f \equiv - g^{\alpha\beta} \nabla_\alpha \nabla_\beta f = - g^{\alpha\beta} (\partial_{\alpha\beta} f - \Gamma^\rho_{\alpha\beta} \partial_\rho f)$$

A system $\{ x^\rho \}$ of local coordinates is caled ***harmonic,*** if the x^ρ are local solutions of (III.2.1). For an arbitrary chart, we introduce the four ***harmonic quantities***

(III.2.3) $$F^\rho = \Delta x^\rho = g^{\alpha\beta} \Gamma^\rho_{\alpha\beta}$$

We also put

(III.2.4) $$F_\rho = g_{\rho\sigma} F^\sigma = g^{\alpha\beta} [\alpha \beta, \rho]$$

The F_ρ and the F^ρ depend on potentials and linearly on their first derivatives

b) Introduce the quantities $L_{\alpha\beta}$ given by

(III.2.5) $$2L_{\alpha\beta} = \partial_\alpha F_\beta + \partial_\beta F_\alpha$$

One has

$$2L_{\alpha\beta} = g^{\lambda\mu} \{ \partial_\alpha [\lambda\mu, \beta] + \partial_\beta [\lambda\mu, \alpha] \}$$

in which appear Christoffel symbol. It follows that

$$2L_{\alpha\beta} = \frac{1}{2} g^{\lambda\mu} (\partial_{\alpha\lambda} g_{\beta\mu} + \partial_{\alpha\mu} g_{\beta\lambda} - \partial_{\alpha\beta} g_{\lambda\mu} + \partial_{\beta\lambda} g_{\alpha\mu} + \partial_{\beta\mu} g_{\alpha\lambda} - \partial_{\alpha\beta} g_{\lambda\mu})$$

that is

(III.2.6) $$2L_{\alpha\beta} = g^{\lambda\mu} (\partial_{\alpha\lambda} g_{\beta\mu} + \partial_{\beta\mu} g_{\alpha\lambda} - \partial_{\alpha\beta} g_{\lambda\mu})$$

The components $R_{\alpha\beta}$ of the Ricci tensor depend also on the potentials and on the first and second derivatives. We will prove that $L_{\alpha\beta}$ occurs in a simple way in the terms of $R_{\alpha\beta}$ containing second derivatives. We have according to (III.1.6)

$$R_{\alpha\beta} \simeq -\frac{1}{2} g^{\lambda\mu} \partial_{\lambda\mu} g_{\alpha\beta} + \frac{1}{2} g^{\lambda\mu} (\partial_{\alpha\lambda} g_{\beta\mu} + \partial_{\beta\mu} g_{\alpha\lambda} - \partial_{\alpha\beta} g_{\lambda\mu})$$

where the symbol $\simeq$ signifies modulo terms depending one the potentials and quadratically on their first derivatives. By comparison with (III.2.6), we obtain

$$R_{\alpha\beta} \simeq -\frac{1}{2} g^{\lambda\mu} \partial_{\lambda\mu} g_{\alpha\beta} + L_{\alpha\beta}$$

we state

Theorem. *In an arbitrary chart, each component of the Ricci tensor may be written*

(III.2.7) $$R_{\alpha\beta} = R^{(h)}_{\alpha\beta} + L_{\alpha\beta}$$

where the $L_{\alpha\beta}$ *are defined by formula* (III.2.5) *and where*

(III.2.8) $$R^{(h)}_{\alpha\beta} = -\frac{1}{2}\,\partial_{\lambda\mu}\, g_{\alpha\beta} + A_{\alpha\beta}$$

the $A_{\alpha\beta}$ *depending on the potentials and quadratically on their first derivatives.*

If the chart is harmonic, $L_{\alpha\beta} = 0$ and $R_{\alpha\beta} = R^{(h)}_{\alpha\beta} \simeq -\frac{1}{2} g^{\lambda\mu}\partial_{\lambda\mu}\, g_{\alpha\beta}$.

There is a separation of potentials in terms of their second derivatives.

III.3. - Einstein's equations and harmonic charts.

a) Consider the genral system of Einstein equations, where T is assumed to be of class C^1

(III.3.1) $$S_{\alpha\beta} = x\, T_{\alpha\beta}$$

It follows from the conservation identities (III.1.2) that the system (III.3.1) implies

(III.3.2) $$\nabla_\alpha T^\alpha_\beta = 0$$

If a solution of the Einstein equations is written in a harmonic chart, one has $L_{\alpha\beta} = 0$ and this solution satisfies

(III.3.3) $$S^{(h)}_{\alpha\beta} = x\, T_{\alpha\beta}$$

where

$$S^{(h)}_{\alpha\beta} = R^{(h)}_{\alpha\beta} - \frac{1}{2} R^{(h)} g_{\alpha\beta} \qquad R^{(h)} = g^{\alpha\beta} R^{(h)}_{\alpha\beta}$$

b) Conversely, let Σ be a ***spatial hypersurface,*** $\{ y^\alpha \}$ a chart adapted to Σ . Consider a solution of system (III.3.2), (III.3.3) corresponding to Cauchy data such as for the chart $\{ y \}^\alpha$

(III.3.4) $$F_\rho = 0 \qquad \text{for} \quad y^0 = 0$$

and

(III.3.5) $$S^0_\alpha = \chi T^0_\alpha \qquad \text{for} \quad y^0 = 0$$

We will prove that one also has

(III.3.6) $$\partial_0 F_\rho = 0 \qquad \text{for} \quad y^0 = 0$$

In fact, one has

$$S^0_\alpha = S^{(h)\,0}_\alpha + L^0_\alpha - \frac{1}{2} g^0_\alpha L \qquad L = g^{\lambda\mu} L_{\lambda\mu} = g^{\lambda\mu} \partial_\lambda F_\mu$$

One deduces from (III.3.5) that on the hypersurface Σ one has

(II.3.7) $$L^0_\alpha - \frac{1}{2} g^0_\alpha L = \chi T^0_\alpha - \chi T^0_\alpha = 0$$

On the other hand, according to (III.2.5) and (III.3.4)

$$2 L^0_\alpha = g^{00} \partial_0 F_\alpha + g^0_\alpha g^{0\rho} \partial_0 F_\rho \qquad L = g^{0\rho} \partial_0 F_\rho$$

Equation (III.3.7) gives $g^{00}\, \partial_0 F_\alpha = 0$ over Σ ; g^{00} being $\neq 0$, one has (III.3.6)

c) For our solution of (III.3.2), (III.3.3), we have

$$\nabla_\lambda (L^\lambda_\mu - \frac{1}{2}\, g^\lambda_\mu L) = \nabla_\lambda S^\lambda_\mu - \nabla_\lambda S^{(h)\lambda}_{\ \ \mu} = - \chi \nabla_\lambda\, T^\lambda_\mu$$

and according to (III.3.2)

$$\nabla_\lambda (L^\lambda_\mu - \frac{1}{2}\, g^\lambda_\mu L) = 0 \qquad \text{(III.3.8)}$$

But one has

$$2L^\lambda_\mu - g^\lambda_\mu L = g^{\lambda\rho} (\partial_\rho F_\mu + \partial_\mu F_\rho) - g^\lambda_\mu\, g^{\rho\sigma}\, \partial_\rho F_\sigma$$

and as consequence

$$\partial_\lambda (2L^\lambda_\mu - g^\lambda_\mu L) = g^{\lambda\rho}\, \partial_{\lambda\rho} F_\mu + g^{\lambda\rho}\, \partial_{\lambda\mu} F_\rho - g^{\rho\sigma}\, \partial_{\mu\rho} F_\sigma + U_\mu$$

where U_μ represents terms linear in $\partial_\lambda F_\mu$. In then follows that (III.3.8) may be written

$$g^{\lambda\rho}\, \partial_{\lambda\rho} F_\mu + A^{\alpha\beta}_\mu\, \partial_\alpha F_\beta = 0 \qquad \text{(III.3.9)}$$

where the A are regular functions of potentials and of their first derivatives.

One sees that the F_ϱ satisfy the hyperbolic linear system (III.3.9), which admits a uniqueness theorem since Σ is spatial ; the only solution of (III.3.9) that satisfies $F_\varrho = 0$, $\partial_0 F_\varrho = 0$ over Σ is the zero solution. Thus, $F_\varrho = 0$ outside Σ and for our solution of (III.3.2), (III.3.3) the chart is harmonic. It then follows that this solution is also a solution of the Einstein system (III.3.1).

Theorem.- *Any solution of* (III.3.2), (III.3.3) *that satisfies* $F_\varrho = 0$ *and* $S^0_\alpha = \chi T^0_\alpha$ *over any spatial hypersurface* $\Sigma\,(y^0 = 0)$ *is a solution of the corresponding Cauchy problem of the Einstein system and the corresponding chart is harmonic for this solution.*

III.4. - Strictly hyperbolic matrices.

A fundamental tool for establishing existence and uniqueness theorems relative to mathematical physics over a curved space-time is an important theorem due to Leray concerning the local Cauchy problem for a class of differential systems. We begin by giving the necessary definitions and terms.

a) Let V_{n+1} be a differentiable manifold of class C^{h+1}, where h is sufficiently large. If Ω is a domain of V_{n+1}, $a(x, \partial)$ is a differential operator on the functions defined over Ω, where $x \in \Omega$ and where $\partial = \{\partial_\alpha\}$ $(\alpha = 0, 1, \dots, n)$ for a chart $\{x^\alpha\}$ with domain Ω. We denote by T^*_x the space cotangent at x to V_{n+1}.

If m is the order of **a**, introduce the polynomial $a(x, \xi)$ of degree m in ξ (where $\xi \in T^*_x$) and its principal part $h(x, \xi)$ defined by the set of homogeneous terms of degree m. We denote by $V_x(h)$ the cone of T^*_x defined by the equation

$$h(x, \xi) = 0$$

This cone is considerd as a projective hypersurface. The operator is said to be strictly hyperbolic at x if it satisfies the following assumption :

There exists in T^*_x ***elements*** ξ ***such that each straight line issuing from*** ξ ***and not passing through the vertex cuts the surface of the cone*** $V_x(h)$ ***at*** m ***real and distinct points. Under this condition, the set of these points*** ξ ***is the interior of opposite half cones*** Γ^+_x (a) and Γ^-_x (a) ***whose boundaries belong to*** $V_x(h)$.

b) Consider now a diagonal matrix $A(x, \partial)$ sufficiently differentiable in x over Ω

$$A(x,\partial) = \begin{pmatrix} a_1(x_1, \partial) & \ldots\ldots & 0 \\ & & \\ 0 & & a_p(x, \partial) \end{pmatrix} \qquad \text{(III.4.1)}$$

The element $a_\tau(x, \partial)$ ($\tau = 1, \ldots, p$) is a differentiable operator of order $m(\tau)$. The matrix $A(x, \partial)$ is called ***strictly hyperbolic at*** x ***if***

1) *The* a_τ *are strictly hyperbolic at* x

2) *The two opposite convex halfcones*

$$\Gamma_x^+(A) = \bigcap_\tau \Gamma_x^+(a_\tau) \qquad \Gamma_x^-(A) = \bigcap_\tau \Gamma_x^-(a_\tau)$$

have a non empty interior.

c) Introduce in the tangent space T_x at x to V_{n+1}, the convex halfcone $C_x^+(A)$ dual to $\Gamma_x^+(A)$. This is the set of vectors V of T_x such that $\xi(V) \geqslant 0$ for all $\xi \in \Gamma_x^+(A)$. Similarly the opposite halfcone $C_x^-(A)$ is dual to $\Gamma_x^-(A)$ and we put $C_x(A) = C_x^+(A) \cup C_x^+(A)$.

A path of V_{n+1} is said to be ***temporal for*** A if the halftangent at each point x of the path belongs to $C_x^+(A)$. A regular hypersurface Σ of V_{n+1} is called ***spatial for*** A, if the tangent subspace at $x \in \Sigma$ to Σ is exterior to $C_x(A)$.

The matrix $A(x,\partial)$ is called ***strictly hyperbolic over*** Ω if the two following conditions are satisfied

1) $A(x, \partial)$ *is strictly hyperbolic at each point of* Ω

2) *The set of temporal paths for* A *joining in* Ω *two points* x, x' *of* Ω *is compact or empty in the space of paths equipped with the compact-open topology.*

It then follows that over Σ, the set of corresponding points is compact or empty. If $A(x, \partial)$ is strictly hyperbolic at x, one shows that there exists a contractile domain Ω containing x such that $A(x, \partial)$ is strictly hyperbolic over Ω. We choose in the following such a domain.

III.5. - Leray systems.

a) Consider over the differentiable manifold V_{n+1} a quasilinear differential system in unknowns $U = (u_\sigma)$ $(\sigma = 1, \dots, p)$ of the form :

(III.5.1) $$A(x, U, \partial) = B(x, U)$$

where A is a diagonal matrix $A(x, U, \partial) = (a_\tau(x, u, \partial)$ $(\tau = 1, \dots, p)$, the a_τ being differential operators of respective orders $m(\tau)$ and where $B(x, U) = (b_\tau(x, U))$.

In order to make precise the hypothesis that we are going to make on the a_τ and b_τ, we associate to each unknown u_σ an integral index $s(\sigma) \geqslant 1$ and to each equation τ of the system an integral index $t(\tau) \geqslant 1$ such a way that

$$m(\tau) = s(\tau) - t(\tau) + 1$$

We suppose that the b_τ and the coefficients of a_τ are sufficiently regular fonctions of x, of u_σ and of the derivatives of u_σ of order $\leqslant s(\tau) - t(\tau)$. If $s(\sigma) - t(\tau) < 0$, the right members b_τ and the coefficients of a_τ are then independant of u_σ.

The structure of the system (III.5.1) defines the indices $s(\sigma)$ and $t(\tau)$ up to an additive constant. If the preceding hypothesis are satisfied, we say that (III.5.1) is a ***Leray system.***

b) We now define a Cauchy problem for the system (III.5.1). Let Σ be a local regular hypersurface in Ω. In the neighborhood of Σ we introduce

$$V(x) = (v_\sigma(x)) \qquad (\sigma = 1, \ldots, p)$$

where v_σ has, locally square integrable derivatives of order $\leqslant s(\sigma) + 1$.

We suppose that

1) ***the hypersurface Σ is spatial for*** $A(x, V(x), \partial$ ***where*** $x \in \Sigma$
2) ***the matrix*** $A(x, V(x), \partial)$ ***is strictly hyperbolic over*** Ω
3) ***the differences*** $a_\tau(x, V(x), \partial)\, \partial_\tau - b_\tau(x, V(x))$ ***and their derivatives of order*** $< t(\tau)$ ***are zero over*** Σ.

We consider, under these assumptions, the Cauchy problem for system (III.5.1) defined by the Cauchy data determined by $V(x)$. A solution of this Cauchy problem is a solution $U = \{u_\sigma\}$ of (III.5.1) admitting ***locally integrable derivatives of order*** $\leqslant s(\sigma)$ ***and such that***

$$u_\sigma(x) - v_\sigma(x)$$

and their derivatives of order $< s(\sigma)$ ***are zero*** Σ. For this problem, Leray proved the following existence and uniqueness theorem for a strictly hyperbolic Leray system.

Theorem (Leray). - *Under the above hypothesis, the Cauchy problem for the system* (III.5.1) *admits a solution at least in the neighborhood of a point* x *of* Σ . *If* $U = (u_\sigma)$ *and* $U' = (u'_\sigma)$ *are two solutions of the Cauchy problem in the neighborhood of* x *and if* u_σ *and* u'_σ *have locally square integrable derivatives of order* $\leqslant s(\sigma) + 1$, *these two solutions coincide.*

Thus consider a strictly hyperbolic Leray system. In the usual way, the Cauchy data are here the values over Σ of u_σ and of their derivatives of order $\leqslant s(\sigma) - 1$, subject to suitable compatobility conditions. Leray's theorem says that, for such a system, existence and uniqueness hold for the local Cauchy problem.

III.6. - The Einstein equations in the exterior case.

We will give now relatively simple examples of the use of Leray's theorem, that seems perhaps too abstract at this moment

a) Let us consider the Einstein system for the exterior case, that may be written

(III.6.1) $$R_{\alpha\beta} = 0$$

In harmonic coordinates (III.61) may be written

(III.6.2) $$R^{(h)}_{\alpha\beta} = -\frac{1}{2} g^{\lambda\mu} \partial_{\lambda\mu} g_{\alpha\beta} + f_{\alpha\beta} (g_{\lambda\mu}, \partial_\rho g_{\lambda\mu}) = 0$$

where $f_{\alpha\beta}$ are given regular functions, that is

(III.6.3) $$g^{\lambda\mu}\, \partial_{\lambda\mu}\, g_{\alpha\beta} = 2\, f_{\alpha\beta}\, (g_{\lambda\mu}\, ,\, \partial_\rho\, g_{\lambda\mu}\,)$$

We study the system (III.6.3). It is easy to see that this system is a ***Leray system.*** The matrix A is here the 10 x 10 diagonal matrix whose diagonal elements are all equal to

(III.6.4) $$g^{\lambda\mu}\, \partial_{\lambda\mu}$$

(III.6.2) or (III.6.3) satisfies our differentiability hypothesis with the following indices

$$s(g_{\alpha\beta}\,) = 2 \qquad\qquad t = (R^{(h)}_{\alpha\,\beta}\,) = 1$$

In fact (III.6.4) is of order $2 - 1 + 1 = 2$. The coefficients of (III.6.4) and the right members depend only on the $g_{\lambda\mu}$ and on their derivatives of order at most $2-1 = 1$.

b) This being granted, let Σ be a locally regular hypersurface, $\{ y^\alpha \}$ a chart adaptated to Σ . Consider Cauchy data such that

1) *The quadratic form* $g_{\lambda\mu}\, v^\lambda\, v^\mu$ *is strictly hyperbolic and the hypersurface* Σ *is spatial at each point for the corresponding elementary cone.*

2) *One has over* Σ

$$F_\rho = 0 \qquad\qquad S^0_\alpha = 0 \qquad\qquad (\text{for } y^0 = 0\,)$$

If C_x^+ is the future half-cone, its dual is the half-cone Γ_x^+ defined by

$$(\text{III.6.5}) \qquad g^{\lambda\mu}\,\xi_\lambda\,\xi_\mu \geqslant 0$$

with $\xi(v) \geqslant 0$ if $v \in C_x^+$. It is clear that there is ***strict hyperbolicity.***

Thus, if the Cauchy data are sufficiently regular, it follows from the Leray theorem that the Cauchy problem corresponding to (III.6.2) admits a unique solution.

In 1950 Mrs Choquet-Bruhat established this result by direct means.

The set of temporal paths in Ω issuing from a point y (resp. terminating at this point) in the future defines the future $\mathcal{E}^+(y)$ (resp. the part $\mathcal{E}^-(y)$) of the point y . If $\mathcal{E}(y) = \mathcal{E}^+(y) \cup \mathcal{E}^-(y)$ is the "emission" of y , the boundary of this emission is the ***characteristic conoid*** C_y corresponding at y , set of the isotropic geodesics issuing from this point. It follows from the Leray theory that the values of the solution at a point y of a neighborhood of Σ on the future side depend only on the Cauchy data over $\mathcal{E}^-(y) \cap \Sigma$ (***domain of dependence).***

c) T being zero, it follows from Section (III.3) that the preceding solution of system (III.6.2) everywhere satisfies the harmonicity conditions $F_\varrho = 0$ and is thus ***a solution of the Einstein system*** (III.6.1) . For Cauchy data satisfying $F_\varrho = 0$ over Σ , one thus obtains

the existence of a solution of the Cauchy problem for the Einstein equations in the exterior case.

The uniqueness of the Cauchy problem for the Einstein system is ***a physical or geometrical uniqueness,*** that is uniqueness modulo a change of chart leaving invariant the numerical values of the coordinates at each point of Σ and the Cauchy data at that point.

Consider a solution of the Cauchy problem for the Einstein system $R_{\alpha\beta} = 0$ corresponding to data such as $F_\rho = 0$ over Σ. One may construct a harmonic chart for this solution such that the coordinates of each point of Σ and the Cauchy data keep the same values. To that end, consider the system

$$\Delta f^\rho = 0 \qquad \text{(III.6.6)}$$

and the solution of (III.6.6) corresponding to Cauchy data over Σ

$$f^\rho = y^\rho \qquad\qquad \partial_\alpha f^\rho = \delta^\rho_\alpha$$

The existence theorem for (III.6.6) implies the existence of such functions f^ρ , which define a harmonic chart. One sees easily that, since $F^\rho = 0$ over Σ , the Cauchy data of the initial problem maintain the same values in the harmonic coordinates $\{ f^\rho \}$.

We see thus that, if $F_\rho = 0$ over Σ , any solution of the Cauchy problem relative to $R_{\alpha\beta} = 0$ may be deduced from the unique solution of the same problem for $R^{(h)}_{\alpha\beta} = 0$ by a change of chart leaving invariant the numerical values of the coordinates of the points of Σ and of the Cauchy data.

We have thus obtained existence and uniqueness theorems for the local Cauchy problem relative to Einstein equations in the exterior case.

III.7. - The case of pure matter.

We propose to study the system of Einstein equations corresponding to the case where the only sources are of pure matter. The energy tensor T may be written

$$T_{\alpha\beta} = c^2 r \, u_\alpha u_\beta \qquad (r > 0) \qquad (\text{with } \rho = c^2 r)$$

where r is the density of matter and u_α the unitary velocity vector ; u_α being unitary, the dynamic equations

$$\nabla_\alpha T^\alpha_\beta = 0$$

are equivalent to a system formed by the equation of conservation of matter

$$\nabla_\alpha (r \, u^\alpha) = 0$$

and the equations of motion

$$u^\alpha \nabla_\alpha u^\beta = 0$$

which define the lines of flow which are the integral curves of u^α. These are strictly temporal geodesics.

a) Let Σ be a ***spatial*** hypersurface, $\{ y^\alpha \}$ a chart adapted to Σ. The Cauchy data are defined for us by the values of the potentials and of the first derivatives $\partial_0 g_{\alpha\beta}$. From these data, we can deduce the values of S^0_α over Σ. Since

$$S^0_\alpha = \chi \varrho u^0 u_\alpha$$

there results from the unitary character of u_α

$$(\chi \varrho u^0)^2 = g^{\alpha\beta} S^0_\alpha S^0_\beta$$

It is necessary that ***the right-hand side be strictly positive.*** We set

$$(\Omega^0)^2 = g^{\alpha\beta} S^0_\alpha S^0_\beta$$

so that

$$\chi \varrho u^0 = \Omega^0$$

It follows that

$$u_\alpha = S^0_\alpha / \Omega^0 \qquad u^0 = S^{00}/\Omega^0 \qquad \chi \varrho = \frac{(\Omega^0)^2}{S^{00}}$$

where S^{00} ***is assumed to be strictly positive***

b) Consider the Einstein system for the case of pure matter

$$S_{\alpha\beta} = \chi \varrho u_\alpha u_\beta$$

where

(III.7.1) $$g^{\alpha\beta} u_\alpha u_\beta = 1$$

This system may be written

(III.7.2) $$R_{\alpha\beta} = \chi \varrho (u_\alpha u_\beta - \frac{1}{2} g_{\alpha\beta})$$

with (III.7.1). According to our study of Section (III.3), we introduce the system

(III.7.3) $$R^{(h)}_{\alpha\beta} = \chi \varrho (u_\alpha u_\beta - \frac{1}{2} g_{\alpha\beta})$$

with

(III.7.4) $$\nabla_\alpha (\varrho u^\alpha) \equiv u^\alpha \partial_\alpha \varrho + \varrho \nabla_\alpha u^\alpha = 0$$

and

(III.7.5) $$u^\alpha \nabla_\alpha u^\beta = 0$$

It is not possible for the system (III.7.3), (III.7.4), (III.7.5) to choose indices s and t such that the hypothesis of Leray are satisfied. To obtain a strictly hyperbolic Leray system, we substitute for (III.7.3) the derivatives of these equations along the lines of flow and then use (III.7.4), (III.7.5). One has :

$$u^\gamma \nabla_\gamma [\varrho(u_\alpha u_\beta - \frac{1}{2} g_{\alpha\beta})] = u^\gamma \partial_\gamma \varrho(u_\alpha u_\beta - \frac{1}{2} g_{\alpha\beta}) +$$

$$+ \varrho u^\gamma \nabla_\gamma (u_\alpha u_\beta - \frac{1}{2} g_{\alpha\beta})$$

It follows from (III.7.5) that the last term of the right-hand side is zero and thus according to (III.7.4), one obtains :

$$u^\gamma \nabla_\gamma [\varrho(u_\alpha u_\beta - \frac{1}{2} g_{\alpha\beta})] = - \varrho \nabla_\gamma u^\gamma (u_\alpha u_\beta - \frac{1}{2} g_{\alpha\beta})$$

We may replace the ten equations (III.7.3) be the ten equations

$$u^\gamma \nabla_\gamma R^{(h)}_{\alpha\beta} = - \chi \varrho \nabla_\gamma u^\gamma (u_\alpha u_\beta - \frac{1}{2} g_{\alpha\beta})$$

that is

$$\text{(III.7.6)} \quad -\frac{1}{2} u^\gamma g^{\lambda\mu} \partial_{\gamma\lambda\mu} g_{\alpha\beta} = - \chi \varrho \nabla_\gamma u^\gamma (u_\alpha u_\beta - \frac{1}{2} g_{\alpha\beta}) + u^\gamma f_{\alpha\beta\gamma}$$

where $f_{\alpha\beta\gamma}$ depends on the potentials and on their first and second derivatives.

c) We consider now u^α as an arbitrary vector. The unknown functions are the 10 $g_{\alpha\beta}$, the 4 u^α and ϱ. We have thus 15 unknowns. The system

(III.7.6), (III.7.4) (III.7.5) is a system of 15 equations and it is a Leray system.

The matrix A , here 15 x 15 , is a diagonal matrix. The diagonal elements which are equal and which come from the same partial system are denoted by the same symbol (a(6)) corresponds to the partial system (III.7.6)). These diagonal elements are respectively

$$\text{(III.7.7)} \qquad a(6) = -\frac{1}{2} u^\gamma g^{\lambda\mu} \partial_{\gamma\lambda\mu} \qquad\qquad a(4) = a(5) = u^\alpha \partial_\alpha$$

This system satisfies our assumptions on the derivatives with the following indices for the unknowns

$$s(g_{\alpha\beta}) = 3 \qquad s(\rho) = 1 \qquad s(u^\alpha) = 2$$

and for the equations

$$t(6) = 1 \qquad t(4) = 1 \qquad t(5) = 2$$

In fact, a(6) is of order $3 - 1 + 1 = 3$, $a(4) = a(5)$ are of order 1 . For the operator coefficients and the right-hand sides, we give a table showing, for each unknown and equation, the maximal order of derivation appearing according to the chosen indices. A negative order indicates the absence of an unknown in the corresponding equation. One has

$$(6)\begin{cases} g_{\alpha\beta} : 2 \\ \rho : 0 \\ u^\alpha : 1 \end{cases} \qquad (4)\begin{cases} g_{\alpha\beta} : 2 \\ \rho : 0 \\ u^\alpha : 1 \end{cases} \qquad (5)\begin{cases} g_{\alpha\beta} : 1 \\ \rho : -1 \\ u^\alpha : 0 \end{cases}$$

These maxima orders are compatible with the system (III.7.6), (III.7.4), (III.7.5).

d) *The Cauchy data* $g_{\alpha\beta}$, $\partial_0 g_{\alpha\beta}$ *corresponding to the system* (III.7.3) *are supposed such that*

(1) *The quadratic form* $g_{\lambda\mu} v^\lambda v^\mu$ *is strictly hyperbolic and* Σ *is spatial at each point for the corresponding elementary cone.*

(2) *One has over* Σ

$$F_\rho = 0 \quad \textit{for} \quad y^0 = 0$$

(3) *The relations* $S^0_\alpha = \chi\, \rho\, u^0\, u_\alpha$ *define a positive scalar* ρ *and a unitary vector* u^α *over* Σ . If C^+_x is the elementary half-cone in the future, its dual is the half-cone Γ^+_x defined by

$$g^{\lambda\mu}\, \xi_\lambda\, \xi_\mu \geqslant 0$$

with $\xi(v) \geqslant 0$ if $v \in C^+_x$. To the operator $a(4) = a(5)$ corresponds in T^*_x the "half-cone" $\hat{\Gamma}^+_x$ such that $u^\alpha \xi_\alpha \geqslant 0$. If u^λ is temporal for C^+_x , the intersection $\Gamma^+_x \cap \hat{\Gamma}^+_x$ has a non empty interior, which coincides with the interior of Γ^+_x . One sees that the metrix A corresponding to the system (III.7.6), (III.7.4), (III.7.5) is strictly hyperbolic and that $\Gamma^+_x(A) = \Gamma^+_x$. The temporal or spatial directions for A are the temporal or spatial directions for space-time.

From our initial data $g_{\alpha\beta}$, $\partial_0 g_{\alpha\beta}$, we deduce the values over Σ of the second derivatives of the potentials by means of (III.7.3). The equations (III.7.5) give the values over Σ of the first derivatives of the u^α and (III.7.4) those of $\partial_\alpha \rho$. Thus, for the system (III.7.6), (III.7.4), (III.7.5), we know the values over Σ of the unknowns and of their derivatives of order $\leqslant s(\sigma) - 1$. Our system being ***a strictly hyperbolic Leray system,*** Leray's theorem shows that the considered Cauchy problem for this system admits a unique solution $(g_{\alpha\beta}, \rho, u^\alpha)$.

This solution satisfying (III.7.6) and over Σ, (III.7.3), satisfies (III.7.3) everywhere. More over we deduce from (III.7.5)

$$u^\alpha u^\beta \nabla_\alpha u_\beta = \frac{1}{2} u^\alpha \partial_\alpha (u^\beta u_\beta) = 0$$

The vector u^α being unitary over Σ is everywhere unitary.

It follows from Section (III.3) that our solution $(g_{\alpha\beta}, \rho, u^\alpha)$ satisfies (III.7.2) and $\nabla_\alpha T^\alpha_\beta = 0$ and thus is a solution of the Einstein system (III.7.1), (III.7.2). We have established a local existence and uniqueness theorem for the Cauchy problem of the Einstein system corresponding to the case of pure matter.

IV. - GRAVITATIONAL AND ELECTROMAGNETIC SHOCK WAVES

IV.1. - Gravitational shock waves.

Gravitational shock waves have been studied by Taub, Choquet-Bruhat, as well as the author by means of different methods and under different hypothesis. We shall here use the techniques of tensor distributions (Chap. I), which furnishes the most general results.

a) Let $(V_4, \mathbf{g})$ be a space-time that is here only, by hypothesis, a differentiable manifold of class $(C^1$, piecewise $C^3)$ provided with a strictly hyperbolic metric tensor $\mathbf{g}$ of class $(C^0$, piecewise $C^2)$. We consider a domain Ω, and in Ω a hypersurface of local equation $\varphi = 0$ such that

1) *φ is of class C^1 over Ω and of class C^3 over each domain Ω^+ and Ω^- (notation of Section* (I.3)). *The second and third derivatives of φ are regularly discontinuous at the traverse of Σ.*

2) *The metric tensor g is of class C^0 over Ω and of class C^2 over Ω^-. The first and second derivatives of $g_{\alpha\beta}$ are regularly discontinuous at the traverse of Σ.*

One may still introduce over Ω a chart $\{y^\alpha\}$ adapted to Σ. According to the hypothesis, we have for such coordinates

$$[\partial_i g_{\alpha\beta}] = \partial_i [g_{\alpha\beta}] = 0 \qquad (i, j = 1, 2, 3)$$

and we have to consider only the $\partial_0 g_{\alpha\beta}$. In a manner analogous to that of Section (III.1), one may introduce a change of adapted chart defined by the formula

$$y^{\alpha'} = y^{\alpha} + \frac{1}{2} (y^0)^2 \{ \varphi^{(\alpha)} (y^i) + \varepsilon^{(\alpha)} (y^{\lambda}) \} \tag{IV.1.1}$$

where the $\varepsilon^{(\alpha)}$ converge uniformly to zero, as do their derivatives, when $y^0 \to 0$. The coordinate values of each point x of Σ , the values at x of the potentials $g_{\alpha\beta}$ and the first derivatives $\partial_0 g_{ij}$ of the ***significant potentials*** are invariant under this change. But it is possible, by the choice of the $\varphi^{(\alpha)}$, to create or destroy discontinuities of the first derivatives $\partial_0 g_{\lambda 0}$ of ***nonsignificant potentials*** for Σ , discontinuities that are devoid of physical significance. We are led to the following definition.

Definition. - *A gravitational shock wave is defined by a metric tensor* g *satisfying the preceding hypothesis, solution of the Einstein equations in an appropriate weak sense, and such that the first derivatives of the significant potentials for a hypersurface* Σ *are regularly and effectively discontinuous at the traverse of* Σ *which is called the wave front*

b) Choose a chart with domain Ω compatible with the differentiable structure of V_4 . For each pair (α , β) of indices, we consider the potential $g_{\alpha\beta}$ as a local scalar over Ω ; according to (I.4.7), there exist over Ω scalar distributions $\delta g_{\alpha\beta}$ such that

$$\overline{\delta} [\partial_{\gamma} g_{\alpha\beta}] = \ell_{\gamma} \, \delta g_{\alpha\beta}$$

It is equivalent to say (Hadamard conditions) that there exist over Σ quantities $b_{\alpha\beta}$ such that

$$(IV.1.2) \qquad [\partial_\gamma g_{\alpha\beta}] = \ell_\gamma b_{\alpha\beta}$$

with

$$\delta g_{\alpha\beta} = \bar{\delta} b_{\alpha\beta}$$

The quantities $b_{\alpha\beta}$ depend on the discontinuities of the first derivatives of the potentials, but also on the choice of the equation of Σ : if $\ell = d\varphi$ is multiplied by k , $b_{\alpha\beta}$ is multiplied by k^{-1}.

We will study how the system of the $\delta g_{\alpha\beta}$ or the system of the $b_{\alpha\beta}$ is modified under a change of chart (C^1 , piecewise C^3). Consider the two charts $\{x^\alpha\}$ and $\{x^{\alpha'}\}$. We set $(A^\alpha_{\lambda'}) = (\partial x^\alpha / \partial x^{\lambda'})$. We have

$$g_{\lambda'\mu'} = A^\alpha_{\lambda'} \; A^\beta_{\mu'} \, g_{\alpha\beta}$$

By differentiation we obtain

$$[\partial_{\nu'} g_{\lambda'\mu'}] = A^\alpha_{\lambda'} A^\beta_{\mu'} A^\gamma_{\nu'} [\partial_\gamma g_{\alpha\beta}] + ([\partial_{\nu'} A^\alpha_{\lambda'}] A^\beta_{\mu'} + A^\alpha_{\lambda'} [\partial_{\nu'} A^\beta_{\mu'}]) g_{\alpha\beta}$$

But it follows from (I.6.7) that there exist over Σ four quantities t^α such that

$$[\partial_{\nu'} A^\alpha_{\lambda'}] = [\partial_{\nu'\lambda'} x^\alpha] = \ell_{\nu'} \ell_{\lambda'} t^\alpha$$

We obtain :

$$\ell_{\nu'} b_{\lambda'\mu'} = A^\alpha_{\lambda'} A^\beta_{\mu'} A^\gamma_{\nu'} \ell_\gamma b_{\alpha\beta} + \ell_{\nu'} (A^\beta_{\mu'} \ell_{\lambda'} t^\alpha + A^\alpha_{\lambda'} \ell_{\mu'} t^\beta) g_{\alpha\beta}$$

If we set $t_\beta = g_{\alpha\beta} t^\alpha$, we have the following formula

(IV.1.3) $$b_{\lambda'\mu'} = A^\alpha_{\lambda'} A^\beta_{\mu'} (b_{\alpha\beta} + \ell_\alpha t_\beta + \ell_\beta t_\alpha) .$$

If the change of map is of type (IV.1 1) that is tangent to the identity change along Σ and admitting discontinuous second derivatives, we have for the $b_{\alpha\beta}$ the change

(IV.1.4) $$b_{\alpha\beta} \to b_{\alpha\beta} + \ell_\alpha t_\beta + \ell_\beta t_\alpha$$

(IV.1.4) defines ***a gravitational gauge change*** for the $b_{\alpha\beta}$, with arbitraries t_α . If we suppose, an instant, that the derivatives of the non significant potentials are continuous at the traverse of Σ , the t_α are necessarily zero and the corresponding admissible charts are defined up to a C^2-change.

If the change of chart is an arbitrary change of class C^2 , the $b_{\alpha\beta}$ are transformed according to the tensor law

(IV.1.5) $$b_{\alpha\beta} \to b_{\lambda'\mu'} = A^\alpha_{\lambda'} A^\beta_{\mu'} b_{\alpha\beta}$$

The formula (IV.1.3) may be considered as resulting from the composition of (IV.1.4) and (IV.1.5).

The $\delta g_{\alpha\beta}$ are the composition of a tensor distribution δg over Ω , for the C^2-changes of charts. The tensor distribution δg is called ***the infinitesimal discontinuity of*** g , for the considered C^2-structure.

IV.2. - Singular double 2-form.

a) A ***double 2-form*** is, by definition, a 4-tensor H admitting the following symmetry properties

(IV.2.1) $$H_{\alpha\beta,\lambda\mu} = -H_{\beta\alpha,\lambda\mu} = -H_{\alpha\beta,\mu\lambda} \qquad H_{\alpha\beta,\lambda\mu} = H_{\lambda\mu,\alpha\beta}$$

Consider at the point x of V_4 a double 2-form $H \neq 0$ for which there exists a vector $\ell \neq 0$ such that

(IV.2.2) $$\ell_\alpha H_{\beta\gamma,\lambda\mu} + \ell_\beta H_{\gamma\alpha,\lambda\mu} + \ell_\gamma H_{\alpha\beta,\lambda\mu} = 0$$

and

(IV.2.3) $$\ell^\alpha H_{\alpha\beta,\lambda\mu} = 0$$

If such is the case, H is said to be ***singular at*** x ***with fundamental vector*** ℓ .

For each pair (λ, μ) of indices, consider the 2-form at x

$$\pi_{(\lambda\mu)} = \frac{1}{2} H_{\alpha\beta,\lambda\mu}\, dx^\alpha \wedge dx^\beta$$

According to (IV.2.2), (IV.2.3), these 2-forms are singular with fundamental vector ℓ , at least one of which is not zero. It follows that ℓ ***is isotropic*** and defined to within $\ell \rightarrow k\ell$. $(k \in \mathbb{R})$

Consider the symmetrical tensor

$$H_{\alpha\beta} = g^{\varrho\sigma} \; H_{\alpha\varrho\,,\,\beta\sigma}$$

It follows from (IV.2.2) by contraction that

$$\text{(IV.2.4)} \qquad \ell_\alpha \, H_{\beta\gamma\,,\,\alpha\mu} = \ell_\beta \, H_{\gamma\mu} - \ell_\gamma \, H_{\beta\mu}$$

According to (IV.2.3) the left-hand side is zero and there exists a scalar τ such that

$$\text{(IV.2.5)} \qquad H_{\alpha\beta} = \tau \, \ell_\alpha \, \ell_\beta$$

Conversely, ***if the double 2-form*** $H \neq 0$ ***satisfies relations*** (IV.2.2) ***and*** (IV.2.5) ***for a vector*** $\ell \neq 0$, ***one deduces from*** (IV.2.4) ***that*** (IV.2.3) ***is satisfied and*** H ***is singular at*** x ; in particular, ℓ is necessarily ***isotropic.***

h) Introduce the same normalized vectors $v_{(1)}$, $v_{(2)}$ orthogonal to ℓ , such that $v^2_{(1)} = v^2_{(2)} = -1$, , $v_{(1)}$, $v_{(2)} = 0$ as in Section (II.2). Since the 2-forms $\pi_{(\lambda\mu)}$ are singular, there exist coefficients $a_{(u)\,\lambda\mu}$ such that

$$H_{\alpha\beta\,,\,\lambda\mu} = \sum_{u,v} a_{(u)\lambda\mu} \, (\ell_\alpha \, v_{(u)\,\beta} - \ell_\beta \, v_{(u)\alpha}) \qquad (u = 1\,,\,2)$$

According to the symmetry properties of H , for each u , $\{ a_{(u)\,\lambda\mu} \}$ defines a singular 2-form and one obtains for a singular double 2-form, the expression

$$\text{(IV.2.6)} \qquad H_{\alpha\beta\,,\,\lambda\mu} = \sum_{u,v} a_{uv} \, (\ell_\alpha \, v_{(u)\,\beta} - \ell_\beta \, v_{(u)\alpha}) \, (\ell_\lambda \, v_{(v)\,\mu} - \ell_\mu \, v_{(v)\,\lambda})$$

where we may suppose $a_{uv} = a_{vu}$. From (IV.2.6) one deduces by contraction

$$(IV.2.7) \qquad H_{\alpha\beta} = -(a_{11} + a_{22})\, \ell_\alpha \ell_\beta \qquad \tau = -(a_{11} + a_{22})$$

Introduce the symmetrical tensor c defined by

$$(IV.2.8) \qquad c_{\alpha\lambda} = \sum_{u,v} a_{uv}\, V_{(u)\alpha}\, V_{(v)\lambda}$$

We see that (IV.2.6) may be written

$$(IV.2.9) \quad H_{\alpha\beta,\lambda\mu} = c_{\alpha\lambda}\ell_\beta \ell_\mu + c_{\beta\mu}\ell_\alpha \ell_\lambda - c_{\alpha\mu}\ell_\beta \ell_\lambda - c_{\beta\lambda}\ell_\alpha \ell_\mu = \hat{\Sigma}\, c_{\alpha\lambda}\ell_\beta \ell_\mu$$

where $\hat{\Sigma}$ represents the corresponding sum of four terms

c) The following lemma is useful.

Lemma. - *Given a vector* $\ell \neq 0$ *, if* u *double 2-form* H *admits in terms of* ℓ *and of a symmetrical 2-tensor* c *the expression* (IV.2.9), *the tensor* c *is defined by* ℓ *up to the transformation*

$$(IV.2.10) \qquad c_{\alpha\lambda} \to c_{\alpha\lambda} + \ell_\alpha t_\lambda + \ell_\lambda t_\alpha$$

where $\{t_\alpha\}$ *is an arbitrary vector*

In fact if H = 0 , we obtain by product of (IV.2.9) by W^β , where w is an arbitrary vector

$$\{(w^\beta \ell_\beta) c_{\alpha\lambda} - w^\beta c_{\beta\lambda} \ell_\alpha\} \ell_\mu - \{(w^\beta \ell_\beta) c_{\alpha\mu} - w^\beta c_{\beta\mu} \ell_\alpha\} \ell_\lambda = 0$$

It follows that there exists a vector a_α such that

$$(w^\beta \ell_\beta) c_{\alpha\lambda} - w^\beta c_{\beta\lambda} \ell_\alpha = a_\alpha \ell_\lambda$$

Suppose w such that $w^\beta \ell_\beta \neq 0$. We may set

$$a_\alpha = (w^\beta \ell_\beta) t_\alpha$$

and we have

$$(w^\beta \ell_\beta) c_{\alpha\lambda} = (w^\beta \ell_\beta) \ell_\lambda t_\alpha + \ell_\alpha w^\beta c_{\beta\lambda}$$

We see that for a suitable vector u , we have

$$c_{\alpha\lambda} = \ell_\lambda t_\alpha + \ell_\alpha u_\lambda = \ell_\lambda t_\alpha + \ell_\alpha t_\lambda + \ell_\alpha (u_\lambda - t_\lambda)$$

According to the symmetry of $c_{\alpha\lambda}$, the vector $(u_\lambda - t_\lambda)$ is colinear to ℓ_λ and the corresponding term proportional to $\ell_\alpha \ell_\lambda$ may be absorbed by a change of the vector t . It is clear that if $c_{\alpha\lambda} = \ell_\alpha t_\lambda + \ell_\lambda t_\alpha$ for an arbitrary vector t , we have $H = 0$. Our lemma is proved.

d) Consider now ***a singular double 2-form*** H and a tensor c associated with H and defined by (IV.2.8) ; c satisfies

$$c_{\alpha\lambda} \ell^\lambda = 0 \qquad \text{(IV.2.11)}$$

According to the lemma, this tensor c is defined up to a transformation (IV.2.10) where t is such that

(IV.2.12) $\ell^\alpha t_\alpha = 0$

It follows that the scalar

$$e = c^{\alpha\beta} c_{\alpha\beta}$$

depends only on the singular double 2-form H and on the choice of the vector ℓ. From (IV.2.8) one has

(IV.2.13) $$e = \sum_{u,v} (a_{uv})^2$$

We see that e is positive and that $e = 0$ implies $H = 0$

IV.3. - Einstein's equations in the weak sense. Conditions of shock.

a) Consider now a ***gravitational shock wave.*** Fix over Ω an admissible chart $\{x^\alpha\}$. If (β, μ) is a given pair of indices, we consider the corresponding coefficients of connection as defining over Ω a local vector $\Gamma^\alpha_{(\beta\mu)}$. The components of the curvature tensor over Ω^+ or Ω^- are :

$$R^\alpha_{\beta,\lambda\mu} = \partial_\lambda \Gamma^\alpha_{\beta\mu} - \partial_\mu \Gamma^\alpha_{\beta\lambda} + \Gamma^\alpha_{\rho\lambda} \Gamma^\rho_{\beta\mu} - \Gamma^\alpha_{\rho\mu} - \Gamma^\rho_{\beta\lambda}$$

We have :

$$\nabla_\lambda \Gamma^\alpha_{\rho\mu} = \partial_\lambda \Gamma^\alpha_{\beta\mu} + \Gamma^\alpha_{\rho\lambda} \Gamma^\rho_{\beta\mu}$$

Therefore the components of the curvature tensor may be written over Ω^+ or Ω^-

$$R^{\alpha}_{\beta\,,\lambda\mu} = \nabla_\lambda \Gamma^{\alpha}_{\beta\mu} - \nabla_\mu \Gamma^{\alpha}_{\beta\lambda}$$

The vector $\Gamma^{\alpha}_{\beta\mu}$ defines over Ω a local vector distribution $(\Gamma^{\alpha}_{\beta\mu})^D$ We introduce systematically ***the curvature tensor distribution*** Q defined by

$$\text{(IV.3.1)} \qquad Q^{\alpha}_{\beta,\lambda\mu} = \nabla_\lambda (\Gamma^{\alpha}_{\beta\mu})^D - \nabla_\mu (\Gamma^{\alpha}_{\beta\mu})^D$$

where the covariant differentiation has been defined as in Section (I.5). According to (I.5.2), we have :

$$\nabla_\lambda (\Gamma^{\alpha}_{\beta\mu})^D = \overline{\delta}\, \ell_\lambda [\Gamma^{\alpha}_{\beta\mu}] + (\nabla_\lambda (\Gamma^{\alpha}_{\beta\mu})^D$$

There results :

$$Q^{\alpha}_{\beta,\lambda\mu} = \overline{\delta}\, [\ell_\lambda \Gamma^{\alpha}_{\beta\mu} - \ell_\mu \Gamma^{\alpha}_{\beta\lambda}] + (R^{\alpha}_{\beta\,,\lambda\mu})^D$$

consider

$$Q_{\alpha\beta,\lambda\mu} = \overline{\delta}\, [\ell_\lambda [\beta\mu,\alpha] - \ell_\mu [\beta\lambda,\alpha]] + (R_{\alpha\beta,\lambda\mu})^D$$

It is easy to see that, according to (IV.1.2)

$$[\ell_\lambda [\beta\mu,\alpha] - \ell_\mu [\beta\lambda,\alpha]] = \frac{1}{2} (b_{\alpha\mu} \ell_\beta \ell_\lambda + b_{\beta\lambda} \ell_\alpha \ell_\mu - b_{\alpha\lambda} \ell_\beta \ell_\mu - b_{\beta\mu} \ell_\alpha \ell_\lambda)$$

We obtain the following proposition.

Proposition. - *The tensor distribution of curvature* Q *is given over* Ω *by the formula*

$$(IV.3.2) \qquad Q_{\alpha\beta,\lambda\mu} = \bar{\delta}\, H_{\alpha\beta,\lambda\mu} + (R_{\alpha\beta,\lambda\mu})^D$$

where H *is the double 2-form defined over* Σ *by*

$$(IV.3.3) \qquad H_{\alpha\beta,\lambda\mu} = -\frac{1}{2}\,\hat{\Sigma}\; b_{\alpha\lambda}\ell_\beta \ell_\mu$$

It is clear from the definition of Q that H is invariant under the gravitational gauge change (IV.1.4) and it is easy to verify directly this invariance. Moreover Q depends only on the hypersurface Σ and on the discontinuities of the first derivatives of the potentials and it is the same for the tensor distribution $\bar{\delta}H$; if one change ℓ into $k\ell$, $\bar{\delta}H$ is invariant.

The Ricci tensor distribution corresponding to Q is given by

$$Q_{\alpha\beta} = \bar{\delta}\, H_{\alpha\beta} + (R_{\alpha\beta})^D$$

where

$$(IV.3.4) \qquad H_{\alpha\beta} = \frac{1}{2}\,\hat{(}\, b_{\alpha\rho}\ell^\rho \ell_\beta + b_{\beta\rho}\ell^\rho \ell_\alpha - b\,\ell_\alpha \ell_\beta - b_{\alpha\beta}\ell^\rho \ell_\rho\,)$$

with

$$b = g^{\alpha\beta}\, b_{\alpha\beta}$$

Introduce for a moment a chart $\{y^\alpha\}$ ***adapted to*** Σ . One has $\ell_i = 0$ and (IV.3.3) gives

$$H_{\alpha\beta,ij} = 0 \qquad\qquad H_{0i,0j} = -\frac{1}{2} b_{ij}$$

We see that it is equivalent to suppose $H \neq 0$ or that the derivatives of the significant potentials for Σ are effectively discontinuous.

b) *We suppose in the following that the energy tensor* T *is continuous over each domain* Ω^+ *or* Ω^- *and regularly discontinuous at the traverse of* Σ .

Put $T' = \mu(T)$ (notation of Section (III.1))

$$\alpha'_{\alpha\beta} = T_{\alpha\beta} - \frac{1}{2} g_{\alpha\beta} T$$

Einstein's equations may be written

(IV.3.5) $$R_{\alpha\beta} = \chi T'_{\alpha\beta}$$

If T'^D *is the tensor distribution defined by* T' , *we adopt as Einstein's equations over* Ω *in a weak sense, the following system*

(IV.3.6) $$Q_{\alpha\beta} = \chi (T'_{\alpha\beta})^D$$

By introducing a test 2-tensor with compact support in Ω' (resp Ω^-) one sees that (IV.3.6) is compatible over Ω^+ (resp. Ω^-) with (IV.3.5). System (IV.3.5) implies that in the sense of distributions, one has over Ω

$$(R_{\alpha\beta})^D = \chi(T'_{\alpha\beta})^D \tag{IV.3.7}$$

By subtraction of (IV.3.6) and (IV.3.7), there results

$$H_{\alpha\beta} = 0 \tag{IV.3.8}$$

which are ***the conditions of shock.***

According to (IV.3.4), the double 2-form satisfies (IV.2.2). But (IV.3.8) is a particular case of (IV.2.5). Thus H ***is singular with fundamental vector*** ℓ at each point of Σ and ℓ ***is isotropic. The hypersurface*** Σ ***is necessarily characteristic*** for the Einstein system.

One sees that the situation differs completely from the corresponding situation for hydrodynamic shock waves (see Chap. V). In this latter case, the ordinary waves propagate at the speed of sound (that which corresponds to the characteristics) and the shock waves are necessarily supersonic before the shock, subsonic after the shock. Here it is the speed c that plays the role of a sonic speed ; heuristically, one may say that the speed of gravitational shock waves cannot be "supersonic" with respect to c and is exactly the speed c . The shock wave front is itself exactly characteristic, as for ordinary waves.

c) H being singular satisfies (IV.2.3) . From (IV.3.3) , it follows

$$b_{\alpha\lambda}\ell^\lambda \ell_\beta - b_{\beta\lambda}\ell_\lambda \ell_\alpha = 0$$

and there exists a scalar σ such that :

$$b_{\alpha\lambda}\,\ell^{\lambda} = \sigma\,\ell_{\alpha}$$

Denote by b' the tensor $\mu(\sigma b \pi)$ associated to b

(IV.3.9) $$b'_{\alpha\beta} = b_{\alpha\beta} - \frac{b}{2}\,g_{\alpha\beta}$$

The conditions of shock (IV.3.4) may be written

(IV.3.10) $$(b'_{\alpha\varrho}\,\ell^{\varrho})\,\ell_{\beta} + (b'_{\beta\varrho}\,\ell^{\varrho})\,\ell_{\alpha} = 0$$

According to (IV.3.9) we have

$$b'_{\alpha\varrho}\,\ell_{\varrho} = (\,\sigma - \frac{b}{2}\,)\,\ell_{\alpha}$$

and (IV.3.10) implies

(IV.3.11) $$b'_{\alpha\varrho}\,\ell^{\varrho} = 0$$

We obtain thus ***the four conditions of shock,*** invariant under (IV.1.4)

(IV.3.12) $$b_{\alpha\varrho}\,\ell^{\varrho} = \frac{b}{2}\,)\,\ell_{\alpha}$$

Under (IV.1.4), the scalar b is transformed according to

$$b \to b + 2\ell^{\varrho}\,t_{\varrho}$$

and it is possible to choose t_{ϱ} so that $b = 0$. If this is done, in a new change of gauge maintaining $b = 0$, one has $\ell^{\varrho} t_{\varrho} = 0$. If we put in this case

$$c_{\alpha\lambda} = -\frac{1}{2}\ )\ b_{\alpha,1}$$

formula (IV.3.3) takes the form (IV.2.9), where the $c_{\alpha\lambda}$ satisfy (IV.2.11) according to the conditions of shock.

d) One may interpret (IV.3.12) in terms of harmonic quantities F_{ϱ} associated to an arbitrary chart. From

$$F_{\varrho} = \frac{1}{2} g^{\lambda\mu} (\partial_{\lambda} g_{\mu\varrho} + \partial_{\mu} g_{\lambda\varrho} - \partial_{\varrho} g_{\lambda\mu})$$

we deduce

$$[F_{\varrho}] = \frac{1}{2} g^{\lambda\mu} (\ell_{\lambda} b_{\mu\varrho} + \ell_{\mu} b_{\lambda\varrho} - \ell_{\varrho} b_{\lambda\mu})$$

that is

$$(IV.3.13) \qquad [F_{\varrho}] = b_{\lambda\varrho} \ell^{\lambda} - \frac{b}{2} \ell_{\varrho}$$

and we see that the conditions (IV.3.12) express the continuity across Σ of the four harmonic quantities F_{ϱ} . One has

Theorem. - *A front of an effective gravitational shock wave is necessarily characteristic and there exist in this case four conditions of shock given by* (IV.3.12) *or by the continuity across* Σ *of the four harmonic quantities* F_{ϱ} *corresponding to an arbitrary admissible chart.*

e) Introduce the scalar, ***invariant under change of gauge,*** $e_{(2)}$ given by

(IV.3.14) $$\chi\, e_{(2)} = \frac{1}{4}\, b^{\alpha\beta}\, b'_{\alpha\beta} = \frac{1}{4}\, (\, b^{\alpha\beta}\, b_{\alpha\beta} - \frac{1}{4}\, b^2\,)$$

For a gauge such that $b = 0$, one has with the preceding (Section (IV.2))-notations

$$\chi\, e_{(2)} = \frac{1}{4}\, b^{\alpha\beta}\, b_{\alpha\beta} = c^{\alpha\beta}\, c_{\alpha\beta} = \sum_{u,v} (a_{uv})^2 \geqslant 0$$

But since $H_{\alpha\beta} = 0$, one has from (IV.2.7)

(IV.3.15) $$a_{11} + a_{22} = 0$$

One sees that our gravitional wave, described by H written according to (IV.2.6) admits two independant modes corresponding to $a_{11} = - a_{22}$ and a_{12}. One has

(IV.3.16) $$\chi\, e_{(2)} = 2\, \{(a_{11})^2 + (a_{12})^2\} = 2\, \{(a_{12})^2 - a_{11}\, a_{22}\} \geqslant 0$$

The scalar $e_{(2)} \geqslant 0$ may be considered as giving the strenght of the gravitational shock at the point on Σ considered.

IV.4. - A geometric consequence.

a) The conditions of shock involve the following proposition

Proposition. - *The covariant differentiation operator* $\ell^\lambda \nabla_\lambda$ *along* ℓ *is well defined over* Σ. *The scalar* $\nabla_\lambda \ell^\lambda$ *is continuous at the traverse of* Σ.

In fact, $\ell^\lambda \nabla_\lambda$ is well defined over Σ if the quantities $\ell^\lambda \Gamma^\alpha_{\beta\lambda}$ are continuous across Σ. It is equivalent to studying the continuity of the quantities $\ell^\lambda [\beta\lambda, \alpha]$. The discontinuities of these quantities are equal to

$$\frac{1}{2} \ell^\lambda (\ell_\lambda b_{\alpha\beta} + \ell_\beta b_{\alpha\lambda} - \ell_\alpha b_{\beta\lambda}) = \frac{1}{2} (b_{\alpha\lambda} \ell^\lambda \cdot \ell_\beta - b_{\beta\lambda} \ell^\lambda \cdot \ell_\alpha)$$

and are zero according to the conditions of shock.

On the other hand, according to our hypothesis, φ is (C^1, piecewise C^2). It follows from (I.6.7) that there exists a scalar m such that

$$[\partial_\lambda \ell_\mu] = [\partial_{\lambda\mu} \varphi] = m \ \ell_\lambda \ell_\mu$$

There results

$$[\nabla_\lambda \ell_\mu] = m \ \ell_\lambda \ell_\mu - [\Gamma^\rho_{\lambda\mu}] \ \ell_\rho = m \ \ell_\lambda \ell_\mu - \ell^\rho [[\lambda\mu, \rho]]$$

But

$$\ell^\rho [[\lambda\mu, \rho]] = \frac{1}{2} \ell^\rho (\ell_\lambda b_{\mu\rho} + \ell_\mu b_{\lambda\rho} - \ell_\rho b_{\lambda\mu}) = \frac{b}{2} \ell_\lambda \ell_\mu$$

and

$$[\nabla_\lambda \ell_\mu] = (m - \frac{b}{2})\, \ell_\lambda \ell_\mu \qquad \text{(IV.4.1)}$$

By contraction, one sees that $\nabla_\lambda \ell^\lambda$ is continuous.

b) It may be deduced from this proposition, as in Section (II.5) that ℓ being isotropic, one has in particular over Σ

$$\ell^\lambda \nabla_\lambda \ell^\mu = 0 \qquad \text{(IV.4.2)}$$

where the first term is well defined over Σ, which is generated by gravitational rays that are isotropic geodesics.

Suppose H written according to (IV.2.6) at $x \in \Sigma$, where $v_{(1)}$ and $v_{(2)}$ are two normalized spatial vectors orthogonal to ℓ and to each other and sio tangential to the hypersurface Σ.

If x' belongs to the same isotropic geodesic as x, we suppose that the vectors v'_1, v'_2 at x' are deduced of the vectors at x by ***parallel transport*** along the isotropic geodesic.

IV.5. - Propagation of discontinuities.

Throughout we suppose that T is regularly discontinuous at the traverse of Σ. We propose to study the propagation of H along the rays.

a) Fix over Ω an admissible chart $\{x^\alpha\}$. Einstein's equations may be written over Ω^+ or Ω^-, according to Section (III.2).

(IV.5.1) $$- g^{\lambda\mu} \partial_{\lambda\mu} g_{\alpha\beta} + \partial_\alpha F_\beta + \partial_\beta F_\alpha + \psi_{\alpha\beta} = 2\chi T'_{\alpha\beta}$$

where $\psi_{\alpha\beta}$ which is quadratic with respect to the first derivatives, may be written

(IV.5.2) $$\psi_{\alpha\beta} = A^{\lambda\mu\nu,\rho\sigma\tau}_{\alpha\beta} \partial_\nu g_{\lambda\mu} \partial_\tau g_{\rho\sigma}$$

where the A depend only on the potentials and are symmetrical with respect to sets of indices $(\lambda\mu\nu)$ and $(\rho\sigma\tau)$. From (IV.5.1) one deduces

(IV.5.3) $$- g^{\mu} \bar{\delta} [\partial_{\lambda\mu} g_{\alpha\beta}] + \bar{\delta} [\partial_\alpha F_\beta + \partial_\beta F_\alpha] + \bar{\delta} [\psi_{\alpha\beta}] = 2\chi \bar{\delta} [T'_{\alpha\beta}]$$

We have

$$[\psi_{\alpha\beta}] = A^{\lambda\mu\nu,\rho\sigma\tau}_{\alpha\beta} [\partial_\nu g_{\lambda\mu} \partial_\tau g_{\rho\sigma}]$$

that is

$$[\psi_{\alpha\beta}] = A^{\lambda\mu\nu,\rho\sigma\tau}_{\alpha\beta} (\partial_\nu g_{\lambda\mu})^+ [\partial_\tau g_{\rho\sigma}] + A^{\lambda\mu\nu,\rho\sigma\tau}_{\alpha\beta} (\partial_\tau g_{\rho\sigma})^- [\partial_\nu g_{\lambda\mu}]$$

It follows

(IV.5.4) $$\bar{\delta} [\psi_{\alpha\beta}] = A^{\lambda\mu\nu,\rho\sigma\tau}_{\alpha\beta} \{ (\partial_\nu g_{\lambda\mu})^+ \ell_\tau (\partial_\nu g_{\lambda\mu})^- \} \delta g_{\rho\sigma}$$

On the other hand, if we consider the $g_{\alpha\beta}$ as local scalars over Ω, we deduce from (I.6.7) that there exist scalar distributions $\bar{g}_{\alpha\beta}$ over Ω such that

$$\bar{\delta} [\partial_{\lambda\mu} g_{\alpha\beta}] = \partial_\lambda \ell_\mu \delta g_{\alpha\beta} + \ell_\lambda \partial_\mu \delta g_{\alpha\beta} + \ell_\mu \partial_\lambda \delta g_{\alpha\beta} + \ell_\lambda \ell_\mu \bar{g}_{\alpha\beta}$$

ℓ being isotropic, it follows by contraction that

$$(\text{IV.5.5}) \qquad g^{\lambda\mu}\,\bar{\delta}\,[\partial_{\lambda\mu} g_{\alpha\beta}] = 2\ell^{\lambda}\,\partial_{\lambda}\,\delta g_{\alpha\beta} + g^{\lambda\mu}\,\partial_{\lambda}\,\ell_{\mu}\,\delta g_{\alpha\beta}$$

Moreover, according to the continuity of the F_β across Σ

$$(\text{IV.5.6}) \qquad \bar{\delta}\,[\partial_\alpha F_\beta] = \ell_\alpha\,\delta F_\beta$$

According to (IV.5.4) – (IV.5.6), Equations (IV.5.3) may be written

$$(\text{IV.5.7}) \qquad \begin{aligned} &2\ell^{\lambda}\,\partial_{\lambda}\,\delta g_{\alpha\beta} + g^{\lambda\mu}\,\partial_{\lambda}\,\ell_{\mu}\,\delta g_{\alpha\beta} - \ell_\alpha\,\delta F_\beta - \ell_\beta\,\delta F_\alpha \\ &- A^{\lambda\mu\nu,\,\rho\sigma\tau}_{\alpha\beta}\,\{(\partial_\nu g_{\lambda\mu})^+ + (\partial_\nu g_{\lambda\mu})^-\}\,\ell_\tau\,\delta g_{\rho\sigma} = -2\chi\,\bar{\delta}\,[T'_{\alpha\beta}] \end{aligned}$$

b) Choose a point x of Σ. In the neighborhood of x, one may define by continuous extension across Σ, $(\Gamma^{\alpha}_{\beta\gamma})^+$ and $(\Gamma^{\alpha}_{\beta\gamma})^-$ and as a consequence the connection of coefficients

$$\bar{\Gamma}^{\alpha}_{\beta\gamma} = \frac{1}{2}\{(\Gamma^{\alpha}_{\beta\gamma})^+ + (\Gamma^{\alpha}_{\beta\gamma})^-\}$$

continuous across Σ. One may suppose by a C^2–change of charts that one has at x

$$\bar{\Gamma}^{\alpha}_{\beta\gamma}(x) = 0$$

We have thus at x

$$\{(\partial_\nu g_{\lambda\mu})^+ + (\partial_\nu g_{\lambda\mu})^-\}_x = 0$$

One has in this chart at x according to (IV.5.7)

$$(\text{IV.5.8}) \qquad 2\ell^{\lambda}\,\bar{\nabla}_{\lambda}\,\delta g_{\alpha\beta} + \nabla_{\lambda}\,\ell^{\lambda}\,\delta g_{\alpha\beta} - \ell_\alpha\,\delta F_\beta - \ell_\beta\,\delta F_\alpha = -2\chi\,\bar{\delta}\,[T'_{\alpha\beta}]$$

where $\overline{\nabla}$ corresponds to $\overline{\Gamma}$. But one knows that $\ell^{\lambda} \underline{\nabla}_{\lambda}$ and $\nabla_{\lambda} \ell^{\lambda}$ are well defined over Σ, so that $\ell^{\lambda} \overline{\nabla}_{\lambda} = \ell^{\lambda} \nabla_{\lambda}$, $\overline{\nabla}_{\lambda} \ell^{\lambda} = \nabla_{\lambda} \ell^{\lambda}$; (IV.5.8) may be written at x

$$\text{(IV.5.9)} \quad 2 \ell^{\lambda} \nabla_{\lambda} \delta g_{\alpha\beta} + \nabla_{\lambda} \ell^{\lambda} \delta g_{\alpha\beta} - \ell_{\alpha} \delta F_{\beta} - \ell_{\beta} \delta F_{\alpha} = -2 \chi \overline{\delta} [T'_{\alpha\beta}]$$

According to the properties of $\overline{\delta}$, (IV.5.9) may be written in terms of $b_{\alpha\beta}$

$$\text{(IV.5.10)} \qquad 2 \ell^{\lambda} \nabla_{\lambda} b_{\alpha\beta} + \nabla_{\lambda} \ell^{\lambda} b_{\alpha\beta} - \ell_{\alpha} \Phi_{\beta} - \ell_{\beta} \Phi_{\alpha} = -2 \chi [T'_{\alpha\beta}]$$

where the Φ_{α} are appropriate quantities over Σ. It follows that (IV.5.9), (IV.5.10) are valid in any admissible chart and at any point of Σ

c) The system (IV.5.10) implies

$$\text{(IV.5.11)} \quad 2 \ell^{\lambda} \nabla_{\lambda} b'_{\alpha\beta} + \nabla_{\lambda} \ell^{\lambda} b'_{\alpha\beta} - \ell_{\alpha} \Phi_{\beta} - \ell_{\beta} \Phi_{\alpha} + g_{\alpha\beta} \ell^{\rho} \Phi_{\rho} = -2 \chi [T_{\alpha\beta}]$$

According to (IV.3.3), the system (IV.5.10) is equivalent to the following differential system involving the tensor H

$$\text{(IV.5.12)} \qquad 2 \ell^{\rho} \nabla_{\rho} H_{\alpha\beta,\lambda\mu} + \nabla_{\rho} \ell^{\rho} H_{\alpha\beta,\lambda\mu} = \chi \hat{\Sigma} [T'_{\alpha\beta}] \ell_{\beta} \ell_{\mu}$$

By multiplying (IV.5.11) by ℓ^{α}, one deduces from the condition of shock $[T_{\alpha\beta}]$ necessarily satisfies

$$\text{(IV.5.13)} \qquad \ell^{\alpha} [T_{\alpha\beta}] = 0$$

a relation that may be expressed from (I.5.2) as

(IV.5.14) $$\nabla_\alpha (T^\alpha_\beta)^0 = 0$$

which expresses the conservation of impulsion-energy.

Suppose that (IV.5.13) is identically satisfied. *It then suffices to verify the conditions of shock* (IV.3.12) *over an initial 2-surface transverse to the rays :* these conditions are then systematically satisfied over Σ as a consequence of (IV.5.11). We have established in particular.

Theorem. - *If the energy tensor is regularly discontinuous at the traverse front* Σ *of a gravitational shock wave, the 4-tensor* H *propagates along the rays generating* Σ *according to the differential system* (IV.5.12). *This system implies* (IV.5.13). *In particular, if the energy tensor is continuous, one has*

(IV.5.15) $$2\ell^\rho \nabla_\rho H_{\alpha\beta,\lambda\mu} + \nabla_\rho \ell^\rho \cdot H_{\alpha\beta,\lambda\mu} = 0$$

choose a chart $\{ y^\alpha \}$ *adapted to* Σ and suppose that $b_{0\alpha} = 0$ According to (IV.3.12) *one has* $b = 0$ and (IV.5.10) may be written

(IV.5.16) $$2\ell^\lambda \nabla_\lambda b_{ij} + \nabla_\lambda \ell^\lambda b_{ij} = -2\chi [T'_{ij}]$$

IV.6. - Consequences.

a) Through multiplying (IV.5.11) by $b^{\alpha\beta}$, one deduces from (IV.3.13) and the conditions of shock

(IV.6.1) $$2\ell^{\lambda} b^{\alpha\beta} \nabla_{\lambda} b'_{\alpha\beta} + \nabla_{\lambda} \ell^{\lambda} b^{\alpha\beta} b'_{\alpha\beta} = -2\chi b^{\alpha\beta} [T_{\alpha\beta}]$$

But one has

$$2 b^{\alpha\beta} \nabla_{\lambda} b'_{\alpha\beta} = 2 b^{\alpha\beta} \nabla_{\lambda} b_{\alpha\beta} - b \nabla_{\lambda} b = \nabla_{\lambda} (b^{\alpha\beta} b_{\alpha\beta} - \frac{1}{2} b^2)$$

We deduce then from (IV.6.1) (see (IV.3.14))

(IV.6.2) $$\nabla_{\rho} (e_{(2)} \ell^{\rho}) = -\frac{1}{2} b^{\alpha\beta} [T_{\alpha\beta}]$$

We shall interpret this relation. Note that

$$\delta g^{\lambda\mu} = - g^{\lambda\alpha} g^{\mu\beta} \delta g_{\alpha\beta} = - \overline{\delta} b^{\lambda\mu}$$

We are led to introduce the tensor of shock energy $\tau^{(2)}$ defined over Σ by :

(IV.6.3) $$\chi \overline{\delta} \tau^{(2)}_{\alpha\beta} = \frac{1}{2} \delta g^{\lambda\mu} H_{\alpha\lambda,\beta\mu}$$

where the tensor

(IV.6.4) $$\chi \tau^{(2)}_{\alpha\beta} = -\frac{1}{2} b^{\lambda\mu} H_{\alpha\lambda,\beta\mu}$$

is without trace, invariant under change of gauge, and depends only on Σ and on the discontinuities of the first derivatives of the potentials. It follows from the conditions of shock that

(IV.6.5) $$\tau^{(2)}_{\alpha\beta} = e_{(2)} \ell_{\alpha} \ell_{\beta}$$

According to $\ell^{\alpha} \nabla_{\alpha} \ell_{\beta} = 0$ and (IV.6.2) one has

(IV.6.6) $$\nabla_{\alpha} \tau^{(2)\alpha}{}_{\beta} = -\frac{1}{2} b^{\rho\sigma} [T_{\rho\sigma}] \ell_{\beta}$$

If the energy tensor is continuous across Σ, the tensor $\tau^{(2)}$ is conservative.

b) We shall analyze the following situation

The energy tensor T *is supposed to be effectively and regularly discontinuous at the traverse of a characteristic hypersurface* Σ *and* g *is a solution of the Einstein equations (eventually in the weak sense) in a neighborhood of* Σ.

Granting these conditions, the first and second derivatives of the significant potentials cannot be continuous across Σ. One may consider Σ as the wave front of a ***gravitational shock wave,*** effective or not. In general, ***the shock wave is effective :*** if it were not, one could deduce from (IV.5.11) that there necessarily exist quantities Φ_{α} such that

(IV.6.7) $$[T_{\alpha\beta}] = \ell_{\alpha} \Phi_{\beta} + \ell_{\beta} \Phi_{\alpha} - g_{\alpha\beta} \ell^{\rho} \Phi_{\rho}$$

Thus if [T] ***does not have the form*** (IV.6.7), ***one certainly has an effective gravitational shock wave***

If [T] has the form (IV.6.7) and if one does not have an effective shock wave, one necessary has an ordinary gravitational wave (discontinuities of the curvature tensor), according to the Einstein system itself. If this is so, one verifies that (IV.6.7) is satisfied and that one also necessarily (IV.5.13) or (IV.5.14).

IV.7. - Electromagnetic shock waves.

The preceding methods and reasoning may be applied to an electromagnetic shock wave whose front may also be or not the wave front of an effective gravitational shock wave.

a) One has for (V_4, g) ***the hypotheses of Section*** (IV.1). We consider a domain Ω and in Ω a regular hypersurface Σ with equation $\varphi = 0$, where φ sqatisfies the hypothesis of Section (IV.1). We will represent the electromagnetic field F by means of a potential vector α. One introduces thus ***a 1-form*** α ***of class*** C^0 ***over*** Ω ***and of class*** C^2 ***over*** Ω^+ ***and*** Ω^-. ***The first and second derivatives of the components of*** α ***in an admissible chart are regularly discontinuous at the traverse of*** Σ.

α is defined up to a change of gauge $\alpha \to \alpha + dS$ where S is a scalar of class (C^1, piecewise C^3) with respect to Ω, Ω^+ and Ω^-.

Choose an admissible chart $\{x^\alpha\}$ of domain Ω. It follows from (I.4.7) that there exists over Σ a local vector b such that :

$$\partial_\lambda \alpha_\mu = \ell_\lambda b_\mu \tag{IV.7.1}$$

If we change ℓ into $k\ell$, b is changed into k^{-1} b. It follows from (I.6.7) that under a change of electromagnetic gauge, one has

$[\partial_{\lambda\mu} S] = \sigma \ell_\lambda U_\mu$. We obtain for b

(IV.7.2) $$b \to b + \sigma \ell$$

From (IV.7.1), one has for $F = d\alpha$

(IV.7.3) $$[F] = [d\alpha] = \ell \wedge b = H$$

Under our hypothesis F defines over Ω a tensor-distribution F^D.

b) ***We suppose in the following that the electric current vector J is continuous over each domain Ω^+ and Ω^- and regularly discontinuous at the traverse of Σ.***

If ε is the operator of codifferentiation, we have over Ω^+ and Ω^- the Maxwell equations

(IV.7.4) $$\varepsilon F = J$$

If J^D is the vector distribution defined by J over Ω, we adopt as Maxwell's equations in a weak sense, the following system

(IV.7.5) $$\varepsilon F^D = J^D$$

System (IV.7.4) implies that in the sense of distributions, one has over Ω

(IV.7.6) $$(\varepsilon F)^D = J^D$$

where

$$\varepsilon F^{D} = -\,\overline{\delta}\; i(\ell)\, H + (\varepsilon F)^{D} \qquad \text{(IV.7.7)}$$

By subtraction of (IV.7.3) and (IV.7.7)

$$\ell \wedge H = 0 \qquad\qquad i(\ell)\; H = 0$$

and the 2-form H ***is singular*** with fundamental vector ℓ ***necessarily isotropic*** at each point of Σ. It follows that (IV.7.7) is equivalent to

$$i(\ell)\, \mathbf{b} = 0 \qquad \text{(IV.7.8)}$$

which is ***the condition of shock.*** The hypersurface Σ is necessarily characteristic for the Maxwell system. We have the same situation as in the gravitational case.

We may given another interpretation of the ***condition of shock.*** It follows from (IV.7.1) that (IV.7.8) expresses that the quantity $f = g^{\lambda\mu}\, \partial_\lambda\, \alpha_\mu$ ***is continuous across*** Σ .

One has

Proposition. - ***An electromagnetic shock wave front*** Σ ***is necessarily characteristic and there exists, in potential vector terms, a condition of shock that is given by*** (IV.7.8) ***or by the continuity across*** Σ ***of the quantity*** f. ***It follows that*** $F_{\lambda\mu}\, \ell^{\lambda}$ ***is continuous across*** Σ .

IV.8. - Propagation of discontinuities for an electromagnetic shock wave.

We will study the propagation of $[F]$ along the rays that generate an electromagnetic shock wave front Σ, which front may also be a front of a gravitational shock wave. The differentiability hypotheses are the hypotheses of Sections (IV.1) and (IV.7)

a) We choose an admissible chart $\{x^\alpha\}$ of domain Ω and we transform Maxwell equations in an appropriate form where appear the continuous quantities F^ϱ (harmonic quantities) and $f = g^{\lambda\mu}\partial_\lambda \alpha_\mu$. One has over Ω^+ or Ω^-

$$g^{\lambda\mu}\nabla_\lambda F_{\lambda\varrho} = g^{\lambda\mu}(\partial_\lambda F_{\mu\varrho} - \Gamma^\sigma_{\mu\lambda}F_{\sigma\varrho} - \Gamma^\sigma_{\varrho\lambda}F_{\mu\sigma}) = -J_\varrho$$

that is :

$$g^{\lambda\mu}(\partial_{\lambda\mu}\alpha_\varrho - \partial_{\lambda\varrho}\alpha_\mu) + F^\sigma F_{\varrho\sigma} - g^{\lambda\mu}\Gamma^\sigma_{\varrho\lambda}F_{\mu\sigma}) = -J_\varrho$$

we obtain

$$g^{\lambda\mu}\partial_{\lambda\mu}\alpha_\varrho - \partial_\varrho(g^{\lambda\mu}\partial_\lambda\alpha_\mu) + F^\sigma F_{\varrho\sigma} + \frac{1}{2}\partial_\varrho g^{\lambda\mu}(\partial_\lambda\alpha_\mu + \partial_\mu\alpha_\lambda) - $$
$$- g^{\lambda\mu}\Gamma^\sigma_{\varrho\lambda}F_{\mu\sigma} = -J_\varrho$$

It follows after subtraction :

$$g^{\lambda\mu}[\partial_{\lambda\mu}\alpha_\varrho] - \ell_\varrho\Phi + F^\sigma[F_{\varrho\sigma}] + \frac{1}{2}[\partial_\varrho g^{\lambda\mu}(\partial_\lambda\alpha_\mu + \partial_\mu\alpha_\lambda)]$$

(IV.8.1)

$$- g^{\lambda\mu}[\Gamma^\sigma_{\varrho\lambda}F_{\mu\sigma}] = -[J_\varrho]$$

where Φ is an appropriate quantity. According to (I.6.7), one has

(IV.8.2) $$g^{\lambda\mu}[\partial_{\lambda\mu}\alpha_\varrho] = 2\,\ell^\lambda\,\partial_\lambda\,b_\varrho + g^{\lambda\mu}\,\partial_\lambda\,\ell_\mu\,b_\varrho$$

Introduce the same connection $\overline{\Gamma}$ as in Section (IV.5)b . We may write over Σ

$$[\Gamma^\sigma_{\varrho\lambda}F_{\mu\sigma}] = \overline{\Gamma}^\sigma_{\varrho\lambda}[F_{\mu\sigma}] + \frac{1}{2}\{(F_{\mu\sigma})^+ + (F_{\mu\sigma})^-\}\,[\Gamma^\sigma_{\varrho\lambda}]$$

where

$$[\Gamma^\sigma_{\varrho\lambda}] = \frac{1}{2}\,(\ell_\lambda\,b^\sigma_\varrho + \ell_\varrho\,b^\sigma_\lambda - \ell^\sigma\,b_{\varrho\lambda})$$

It follows from the continuity of $F_{\mu\sigma}\,\ell^\mu$ or $F^{\mu\sigma}\,\ell_\mu$ that

(IV.8.3) $$g^{\lambda\mu}\,[\Gamma^\sigma_{\varrho\lambda}F_{\mu\sigma}] = g^{\lambda\mu}\,\overline{\Gamma}^\sigma_{\varrho\lambda}[F_{\mu\sigma}] + F^{\mu\sigma}\,\ell_\mu\,b_{\varrho\sigma}$$

On the other hand

$$[\partial_\varrho\,g^{\lambda\mu}(\partial_\lambda\,\alpha_\mu + \partial_\mu\,\alpha_\lambda)] = \frac{1}{2}\{(\partial_\varrho\,g^{\lambda\mu})^+ + (\partial_\varrho\,g^{\lambda\mu})^-\}\,[\,\partial_\lambda\,\alpha_\mu + \partial_\mu\,\alpha_\lambda\,]$$

$$+\frac{1}{2}\{(\partial_\lambda\,\alpha_\mu + \partial_\mu\,\alpha_\lambda)^+ + (\partial_\lambda\,\alpha_\mu + \partial_\mu\,\alpha_\lambda)^-\}\,[\,\partial_\varrho\,g^{\lambda\mu}\,]$$

b) Choose a point x of Σ . We may suppose that in the chart $\{x^\alpha\}$ the coefficients of the connection $\overline{\Gamma}$ are null at x and that we have choosed an electromagnetic gauge such that

$$\{(\partial_\lambda\,\alpha_\mu + \partial_\mu\,\alpha_\lambda)^+ + (\partial_\lambda\,\alpha_\mu + \partial_\mu\,\alpha_\lambda)^-\}_x = 0$$

We may deduce from the continuity of F^σ across Σ that the coefficients of the connection $\bar{\Gamma}$ being null at x , one has $F^\sigma(x) = 0$. An argument similar at this of Section (IV.5) shows that we deduce from (IV.8.2), (IV.8.3) that (IV.8.1) may be written at x

(IV.8.4) $$2\ell^\lambda \nabla_\lambda b_\rho + \nabla_\lambda \ell^\lambda b_\rho - \ell_\rho \Phi = F^{\mu\sigma} \ell_\mu b_{\rho\sigma} - [J_\rho]$$

This formula is valid over Σ for arbitrary charts and for arbitrary gauges. The first term on the right-hand side shows an interaction between the electromagnetic fields and the eventual gravitatonal shock, an interaction that we shall study.

If the conditions of shock (IV.7.8) and (IV.3.12) are satisfied, it follows from (IV.8.4) that :

(IV.8.5) $$\ell^\rho [J_\rho] = 0$$

which expresses that

$$\nabla_\rho (J^\rho)^D = 0$$

System (IV.8.4) is equivalent to the following differential system holding over $[F]$

(IV.8.6) $$2\ell^\rho \nabla_\rho [F_{\lambda\mu}] + \nabla_\rho \ell_\rho [F_{\lambda\mu}] = -F^{\rho\sigma} H_{\lambda\mu,\rho\sigma} - [\ell_\lambda J_\mu - \ell_\mu J_\lambda]$$

Theorem. - *If the electric current is regularly discontinuous at the traverse of the wave front Σ of an electromagnetic shock wave (and eventually gravitational shock wave), the*

discontinuity [F] ***propagates along the rays generating*** Σ ***according to*** (IV.8.6). ***The electric current satisfies*** (IV.8.5).

(IV.8.7) $$e_{(1)} = - b^{\alpha} b_{\alpha}$$

invariant under a change of gauge. It depends only on [F] and the choice of ℓ and may be considered as giving the strenght of the electromagnetic shock at the point of Σ considered. By multiplying (IV.8.4) by b^{ρ}, one has

$$2\ell^{\lambda} b^{\rho} \nabla_{\lambda} b_{\rho} + \nabla_{\lambda} \ell^{\lambda} b^{\rho} b_{\rho} = F^{\mu\sigma} \ell_{\mu} b^{\rho} b_{\rho\sigma} - b^{\rho} [J_{\rho}]$$

This relation may be written :

(IV.8.8) $$\nabla_{\lambda} (e_{(1)} \ell^{\lambda}) = - F^{\mu\sigma} \ell_{\mu} b^{\rho} b_{\rho\sigma} - b^{\rho} [J_{\rho}]$$

We are led to introduce the "energy" tensor of the electromagnetic shock :

(IV.8.9) $$\tau^{(1)}_{\alpha\beta} = - [F_{\alpha\rho}] [F^{\rho}_{\beta}]$$

By introducing in (IV.8.9) the expression for [F], one obtains

(IV.8.10) $$\tau^{(1)}_{\alpha\beta} = e_{(1)} \ell_{\alpha} \ell_{\beta}$$

The tensor $\tau^{(1)}$ is without trace, invariant under gauge change and depends only on [F]. From (IV.8.8) one has

(IV.8.11) $$\nabla_{\alpha} \tau^{(1)\alpha}_{\ \ \ \beta} = - F^{\mu\sigma} \ell_{\mu} b^{\rho} b_{\rho\sigma} \cdot \ell_{\beta} + b^{\rho} [J_{\rho}] \ell_{\beta}$$

In (IV.8.11) also, the interference between gravitational and electromagnetic phenomena is present.

IV.9. - Gravitational shock wave in an electromagnetic field.

a) Consider *an effective gravitational shock wave in an electromagnetic field* F *for which the electric current is continuous.*

If Σ is the wave front, Σ is eventually also an electromagnetic shock wave front, and over Σ one has from (IV.8.4)

$$2\ell^{\lambda} \nabla_{\lambda} b_{\rho} + \nabla_{\lambda} \ell^{\lambda} b_{\rho} - \ell_{\rho} \Phi = F^{\mu\sigma} \ell_{\mu} b_{\rho\sigma} \tag{IV.9.1}$$

b_{ρ} may be collinear with ℓ_{ρ} only if the vector $F^{\mu\sigma} \ell_{\mu} b_{\rho\sigma}$ is collinear with ℓ_{ρ}. We will analyze this case. Introduce for a moment a chart $\{y^{\alpha}\}$ ***adapted to*** Σ $(\ell_0 = 1 , \ell = 0)$. The collinearity may be written

$$F^{0j} b_{ij} = 0 \tag{IV.9.2}$$

Choose a gauge such that $b = 0$. Since

$$\ell^{0} = 0 \quad \ell^{i} b_{0i} = g^{0i} b_{0i} = b/2 = 0$$

one has

$$\ell^{a} b_{ij} = g^{0j} b_{ij} = 0 \tag{IV.9.3}$$

and

(IV.9.4) $g^{ij}\, b_{ij} = 0$

$\{b_{ij}\}$ defines a symmetrical covariant 2-tensor of null trace over the 3-plane π_x tangent at x to Σ along ℓ. The vectors $\{\ell^j\}$ and $\{F^{0j}\}$ are eigenvectors of $\{b_{ij}\}$ in π_x corresponding to the eigenvalue 0. If these vectors are not collinear, one may choose a basis $\{e_{(1)}, e_{(2)}, e_{(3)}\}$ of π_x such that $e_{(1)} = \ell$ and $e_{(2)}$ are eigenvectors corresponding to 0, $e_{(2)}$ and $e_{(3)}$ being two normalized spatial vectors orthogonal to each other. By an appropriate choice of $e_{(0)}$ $(e^2_{(0)} = 1$, the 2-plane (e_0, e_1) orthogonal to the 2 plane $(e_{(2)}, e_{(3)})$. We may obtain a basis $\{e_{(a)}\}$ such that the metric takes the form

$$g_x(v, v) = (v^0)^2 + 2\, v^0 v^1 - (v^2)^2 - (v^{(3)})^2$$

We use now the basis $\{e_{(\alpha)}\}$ and the basis $\{e_{(i)}\}$ of π_x. The vectors $e_{(1)}$ and $e_{(2)}$ being eigenvectors of $\{b_{ij}\}$ corresponding to 0, one has

$$b_{i1} = 0 \qquad b_{i2} = 0 \qquad (i = 1, 2, 3)$$

and thus $g^{33}\, b_{33} = -\, b_{33} = 0$. We see that $b_{ij} = 0$ and our gravitational shock wave is not effective. It follows that one has necessarily in our case

$$F^{\mu\sigma}\, \ell_\mu = k\, \ell^\sigma \qquad (k \in \mathbb{R})$$

or

(IV.9.5) $i(\ell)\, F \wedge \ell = 0$

Proposition. - *An effective gravitational shock wave with front* Σ *in an electromagnetic field* F *such that* $i(\ell)\, F \wedge$

$\ell \neq 0$ necessarily generates an effective electromagnetic shock wave with the same front.

b) We will prove the following lemma

Lemma. - *The Maxwell tensor τ of an electromagnetic field F admitting an electromagnetic shock wave with front Σ and rays defines by ℓ satisfies identically*

$$\ell^{\alpha} [\tau_{\alpha\beta}] = 0 \tag{IV.9.6}$$

From (II.3.7) one has

$$\tau_{\alpha\beta} = \frac{1}{4} g_{\alpha\beta} F_{\lambda\mu} F^{\lambda\mu} - F^{\rho}_{\alpha} F_{\beta\rho}$$

Study

$$[F^{\rho}_{\alpha} F_{\beta\rho}] = (F^{\rho}_{\alpha})^{+} [F_{\beta\rho}] + (F_{\beta\rho})^{-} [F^{\rho}_{\alpha}]$$

We may write

$$[F^{\rho}_{\alpha} F_{\beta\rho}] = (F^{\rho}_{\alpha})^{+} (\ell_{\beta} b_{\rho} - \ell_{\rho} b_{\beta}) + (F_{\beta\rho})^{-} (\ell_{\alpha} b^{\rho} - \ell^{\rho} b_{\alpha}) \tag{IV.9.7}$$

It follows :

$$[F^{\rho}_{\alpha} F_{\beta\rho}] \ell^{\alpha} = F^{\rho}_{\alpha} \ell^{\alpha} \ell_{\beta} b_{\rho}$$

where $F^{\rho}_{\alpha} \ell^{\alpha}$ is continuous across Σ . From (IV.9.7) we deduce by contraction

(IV.9.8) $$[F_{\lambda\mu} F^{\lambda\mu}] = 2 F^{\lambda\mu} (\ell_\lambda b_\mu - \ell_\mu b_\lambda) = 4 F^{\lambda\mu} \ell_\lambda b_\mu$$

We obtain :

$$\ell^\alpha [\tau_{\alpha\beta}] = F^{\lambda\mu} \ell_\lambda b_\mu \cdot \ell_\beta - F^{\lambda\mu} \ell_\lambda b_\mu \cdot \ell_\beta = 0$$

and the formula (IV.9.6) is proved.

IV.10. - Gravitational and electromagnetic shock waves (with Maxwell tensor).

a) *Suppose that* g , F *are a metric and an electromagnetic field* F *satisfying -eventually in the weak sense- the following Einstein-Maxwell equations.*

(IV.10.1) $$S_{\alpha\beta} = \chi \tau_{\alpha\beta} \qquad \nabla_\alpha F^{\alpha\beta} = - J^\beta$$

and dF = 0 *where* τ *is the Maxwell tensor of* F .

Let Σ be the front of a gravitational and electromagnetic shock wave satisfying (IV.10.1). From (IV.8.11) one has

(IV.10.2) $$\nabla_\alpha \tau^{(1)\alpha}{}_\beta = - (F^{\mu\sigma} \ell_\mu b^\rho b_{\rho\sigma}) \cdot \ell_\beta + b^\rho [J_\rho] \cdot \ell_\beta$$

On the other hand, we have according to (IV.6.6)

(IV.10.3) $$\nabla_\alpha \tau^{(2)\alpha}{}_\beta = - \frac{1}{2} b^{\rho\sigma} [\tau_{\rho\sigma}] \cdot \ell_\beta$$

We seek for evaluate $b^{\rho\sigma}[\tau_{\rho\sigma}]$. Introduce a chart adapted to Σ and a gravitational gauge such that $b = 0$. One has

$$b^{\alpha 0} = 0 \qquad\qquad b^{ij}\, g_{ij} = 0$$

According to (IV.9.7), (IV.9.8), we may write

(IV.10.4) $$[\tau_{\rho\sigma}] = g_{ij}\,(F^{\lambda\mu}\,\ell_\lambda\, b_\mu) + (F_i^{\ \lambda}\,\ell_\lambda)\, b_j + (F_j^{\ \lambda}\,\ell_\lambda)\, b_i$$

It follows

$$b^{\rho\sigma}\,[\tau_{\rho\sigma}] = b^{ij}\,[\tau_{ij}] = 2\, F_{i\lambda}\,\ell^\lambda\, b_j\, b^{ij} = -\,2\, F_{\lambda\sigma}\,\ell^\lambda\, b_\rho\, b^{\rho\sigma}$$

Therefore in an arbitrary chart

(IV.10.5) $$\nabla_\alpha\, \tau^{(2)\alpha}_{\ \ \ \ \beta} = [F^{\lambda\sigma}\,\ell_\lambda\, b^\rho\, b_{\rho\sigma})\,\ell_\beta$$

From (IV.10.2), (IV.10.5) one deduces

(IV.10.6) $$\nabla_\alpha\,\{\tau^{(1)\alpha}_{\ \ \ \ \beta} + \tau^{(2)\alpha}_{\ \ \ \ \beta}\} = b^\rho\,[J_\rho]\cdot\ell_\beta$$

In particular *if* J *is continuous the total tensor of shock* $(\tau^{(1)} + \tau^{(2)})$ *is conservative.*

In terms of scalars $e_{(1)}$, $e_{(2)}$, (IV.10.6) may be written

(IV.10.7) $$\nabla_\alpha\,\{(e_{(1)} + e_{(2)})\,\ell^\alpha\} = b^\rho\,[J_\rho]$$

c) Suppose that the electromagnetic shock wave satisfying (IV.10.1) is effective. From Section (IV.6)b , we knows that if the tensor without trace $[\tau_{\alpha\beta}]$ does not have the form (see (IV.6.7))

$$[\tau_{\alpha\beta}] = \ell_\alpha \Phi_\beta + \ell_\beta \Phi_\alpha \tag{IV.10.8}$$

Σ is also the front of an effective gravitational shock wave that may be considered as generated by the electromagnetic shock wave.

Analyse the situation given by (IV.10.8). Introduce for a moment a chart ***adapted to*** Σ $(\ell_0 = 1 , \ell_i = 0)$. (IV.10.8) is translated by

$$[\tau_{ij}] = 0$$

We have $b^0 = 0$ and according to (IV.9.7) , (IV.9.8)

$$b^i [\tau_{ij}] = (F^{\lambda\mu} \ell_\lambda b_\mu b_\mu) b_j + (F^{\mu\lambda} \ell_\lambda b_\mu) b_j + (F_j^{\lambda} \ell_\lambda) b^\mu b_\mu$$

that is

$$b^i [\tau_{ij}] = -(F_j^{\lambda} \ell_\lambda) e_{(i)} \qquad (e_{(i)} > 0)$$

If $[\tau_{ij}] = 0$, one has $b^i [\tau_{ij}] = 0$ and thus

$$F_j^{\lambda} \ell_\lambda = 0 \tag{IV.10.9}$$

Conversely (IV.10.9) implies $[\tau_{ij}] = 0$; (IV.10.9) is equivalent to say that $F^{\lambda\mu} \ell_\mu$ is collinear with ℓ^μ , let

$$i(\ell) F \wedge \ell = 0 \tag{IV.10.10}$$

Proposition - *Consider a metric* g *and an electromagnetic field* F *satisfying (in the sense) the system* (IV.10.1). *If* F *admits an effective electromagnetic shock wave with front* Σ *and rays defined by* ℓ *and if* $i(\ell)\ F \wedge \ell \neq 0$, *this wave induces a gravitational shock wave with the same front.*

Note that the conditions corresponding to (IV.9) and (IV.10) are the same.

V. - RELATIVISTIC HYDRODYNAMICS

V.1. - Equations of relativistic hydrodynamics.

a) Let (V_4, g) be a space-time satisfying the hypotheses of Section (II.1).

In (V_4, g) an ideal fluid is described by the energy tensor

(V.1.1) $$T^{(f)}_{\alpha\beta} = (\rho + p)\, u_\alpha\, u_\beta - p\, g_{\alpha\beta}$$

where ρ is the proper energy density of the fluid, p its pressure and u_α a unitary velocity vector oriented toward the future. The density ρ includes a ***proper density of matter*** and ***an internal energy.*** We set

$$\rho = c^2\, r\,(1 + \varepsilon/c^2) \qquad (r > 0)$$

where r is the density of matter and ε the ***specific internal energy.*** We denote by $V = 1/r$ the specific volume of the fluid. Consider the scalar

(V.1.2) $$\rho + p = c^2 r\left(1 + \frac{\varepsilon}{c^2} + \frac{p}{c^2 r}\right)$$

We set :

(V.1.3) $$i = \varepsilon + \frac{p}{r} = \varepsilon + pV$$

where i is ***the specific enthalpy*** of the fluid. According to (V.1.2) we introduce the ***fluid index***

$$f = 1 + i/c^2 \qquad (f > 1)$$

with these notations (V.1.1) may be written

(V.1.4) $$T^{(f)}_{\alpha\beta} = c^2 r f\, u_\alpha\, u_\beta - p\, g_{\alpha\beta}$$

The proper temperature of the fluid Θ and its specific entropy S may be defined as in classical hydrodynamics by the differential relation

(V.1.5) $$\Theta\, dS = d\varepsilon + p d\mathbf{V} = di - \mathbf{V}dp = c^2 df - \mathbf{V}dp \qquad (\Theta > 0)$$

One then deduces

(V.1.6) $$c^2\, df = \mathbf{V}dp + \Theta\, dS$$

In relativity the thermodynamic variable $\tau = f\mathbf{V}$ (called the ***dynamic volume)*** plays an important role and is most often substituted for the classic specific volume $\mathbf{V}$. It is convenient to adopt p and S as the fundamental thermodynamic variables. In the following, we shall consider either $r = r\ (p, S)$ or more often $\tau = \tau\ (p,S)$ as given functions of p and S defining an ***equation of state*** for the fluid under consideration.

Let Σ be a hypersurface of V_4 with the local equation $\varphi = 0$, where φ is regular. We set again $\ell = d\varphi$. The speed v^Σ of Σ with respect to the fluid, that is with respect to the temporal direction $\mathbf{u}$ is classically given by (see [9] , p. 42–43).

$$(V.1.7) \qquad \frac{(v^{\Sigma})^2}{c^2} = \frac{(u^{\alpha} \ell_{\alpha})^2}{(u^{\alpha} \ell_{\alpha})^2 - \ell^{\alpha} \ell_{\alpha}} = y^{\Sigma}$$

where

$$(u^{\alpha} \ell_{\alpha})^2 - \ell^{\alpha} \ell_{\alpha} = (u^{\alpha} u^{\beta} - g^{\alpha\beta}) \ell_{\alpha} \ell_{\beta} \geqslant 0$$

$(v^{\Sigma}/c) < 1$ if and only if $\ell^{\alpha} \ell_{\alpha} < 0$, that is if Σ is oriented in time.

b) The differential system of relativistic hydrodynamics is obtained with the following considerations ; first, we suppose that the density of matter r (which corresponds to the specific number of particules) is conserved ; if ∇ is the covariant derivation operator

$$(V.1.8) \qquad \nabla_{\alpha} (r u^{\alpha}) = 0$$

The Eisntein system may be written :

$$(V.1.9) \qquad S_{\alpha\beta} = \chi \, T^{(f)}_{\alpha\beta}$$

which implies the equations of relativistic dynamics :

$$(V.1.10) \qquad \nabla_{\alpha} T^{(f)\alpha}_{\ \ \ \ \beta} = 0$$

We may either consider the Einstein equations (V.1.9) and the equations (V.1.8), (V.1.10) for the simultaneous determination of the metric and of the dynamic of the fluid, or consider the metric as given and conserve only the system (V.1.8), (V.1.10).

We will transform the system (V.1.8), (V.1.10) in an equivalent system ; (V.1.10) may be explicitely written

$$(V.1.11) \qquad \nabla_\alpha (c^2 r f u^\alpha) u_\beta + c^2 r f u^\alpha \nabla_\alpha u_\beta - \partial_\beta p = 0$$

Through multiplying (V.1.11) by u^β , one obtains after taking (V.1.6) into account ($c^2 f'_p = v$; $c^2 f'_s = \Theta$) the relation

$$(V.1.12) \qquad \nabla_\alpha (c^2 r f u^\alpha) - u^\alpha \partial_\alpha p = 0$$

let

$$c^2 f \nabla_\alpha (r u^\alpha) + c^2 r u^\alpha \partial_\alpha f - u^\alpha \partial_\alpha p = 0$$

which may be written

$$c^2 f \nabla_\alpha (r u^\alpha) + r \Theta u^\alpha \partial_\alpha S = 0$$

The relation (V.1.8) thus entails the ***equation of adiabatic flow***

$$(V.1.13) \qquad u^\alpha \partial_\alpha S = 0$$

By taking of this result in (V.1.11), we obtain the differential system of ***the lines of flow***

$$(V.1.14) \qquad c^2 r f u^\alpha \nabla_\alpha u^\beta - (g^{\alpha\beta} - u^\alpha u^\beta) \partial_\alpha p = 0$$

The system (V.1.8), (V.1.10) is equivalent to the system (V.1.8), (V.1.13) and (V.1.14).

V.2. - Characteristic hypersurfaces of the hydrodynamic system.

a) It follows from Section (V.1) that we may consider as main equations for the simultaneous determination of the metric and of the dynamics of the fluid the system

(V.2.1) $$S_{\alpha\beta} = \chi\,(c^2 r f\, u_\alpha u_\beta - p\, g_{\alpha\beta})$$

with

(V.2.2) $$g^{\alpha\beta} u_\alpha u_\beta = 1$$

and the equation of adiabatic flow

(V.2.3) $$u^\alpha \partial_\alpha S = 0$$

Study the Cauchy problem for this system. Consider a regular hypersurface Σ and adapted chart $\{ y^\alpha \}$. We give over Σ $(y^0 = 0)$ the values of the potentials $g_{\alpha\beta}$ and of their first derivatives, as well as the value of S .

We suppose that these Cauchy data are such that the quadratic form $g_{\lambda\mu} v^\lambda v^\mu$ is strictly hyperbolic at each point of Σ , that Σ is not tangent to the elementary cone $(g^{00} \neq 0)$ and that the harmonic quantities F^ρ are zero over Σ .

It is well-known that the values of the S^0_α over Σ depend only on the Cauchy data corresponding to the metric. From

$$S_\alpha^0 = \chi\,(c^2\,rf\,u^0\,u_\alpha - p\,g_\alpha^0)$$

and according to (V.2.2), it follows that

$$(\chi\,c^2\,rf\,u^0)^2 = g^{\alpha\beta}\,(S_\alpha^0 + \chi\,p\,g_\alpha^0)\,(S_\beta^0 + \chi\,p\,g_\beta^0)$$

We set

$$[\Omega^0(p)]^2 = g^{\alpha\beta}\,(S_\alpha^0 + \chi\,p\,g_\alpha^0)\,(S_\beta^0 + \chi\,p\,g_\beta^0) > 0$$

where the right-hand side is a function of the thermodynamical variable p. The variable p is supposed positive in the domain which we are considering. It follows that

(V.2.4) $$u_\alpha = \frac{S_\alpha^0 + \chi p\,g_\alpha^0}{\Omega^0(p)} \qquad u^0 = \frac{S^{00} + \chi\,p\,g^{00}}{\Omega^0(p)}$$

with $u^0 \neq 0$ and

(V.2.5) $$\chi\,c^2\,rf = \frac{[\Omega^0(p)]^2}{S^{00} + \chi\,p\,g^{00}}$$

S being known over Σ ; Equation (V.2.5) is a condition on the possible values of p over Σ. This condition may be written

$$F(p\,;y^i) \equiv \chi\,c^2\,rf\,(S^{00} + \chi\,p\,g^{00}) - g^{\alpha\beta}\,(S_\alpha^0 + \chi\,p\,g_\alpha^0)\,(S_\beta^0 + \chi\,p\,g_\beta^0) = 0$$

It is interesting to evaluate the derivative F'_p . It follows from (V.1.6) that

$$\chi^{-1}\,F'_p = (c^2\,f\,r'_p + 1)\,(S^{00} + \chi\,p\,g^{00}) + \chi\,c^2\,rf\,g^{00} - 2\,g^{0\alpha}\,(S_\alpha^0 + \chi\,p\,g_\alpha^0)$$

that is

$$\chi^{-1} F'_p = (c^2 f r'_p - 1)(S^{00} + \chi p g^{00}) + \chi c^2 r f g^{00}$$

But from (V.2.1)

$$S^{00} + \chi p g^{00} = \chi c^2 r f (u^0)^2$$

It follows :

$$\chi^{-1} F'_p = \chi c^2 r f [(c^2 f r'_p - 1)(u^0)^2 + g^{00}]$$

For appropriate Cauchy data, we have a value of p over Σ satisfying $F(p, x^i) = 0$ and such that $F'_p \neq 0$. The values of the u_α over Σ are then known with

$$g^{00} \neq 0 \qquad u^0 \neq 0 \qquad (c^2 f r'_p - 1)(u^0)^2 + g^{00} \neq 0$$

b) According to Section (V.1), substitute for the system (V.2.1) (V.2.2) (V.2.3) the system

(V.2.6) $$R_{\alpha\beta} = \chi \left(c^2 r f u_\alpha u_\beta - \frac{1}{2}(c^2 r f - 2p) g_{\alpha\beta}\right)$$

the equation (V.2.3) let

(V.2.7) $$u^\alpha \partial_\alpha S = 0$$

the equation

(V.2.8) $$\nabla_\alpha (r u^\alpha) = u^\alpha \partial_\alpha r + r \nabla_\alpha u^\alpha = 0$$

and the system of the lines of flow

(V.2.9) $$c^2 r f u^\alpha \nabla_\alpha u^\beta - (g^{\alpha\beta} - u^\alpha u^\beta) \partial_\alpha p = 0$$

We note that, according to (V.2.9) $u^\alpha u^\beta \nabla_\alpha u_\beta \simeq 0$ modulo $u^\beta u_\beta = 1$; u_β that is initially unitary remains unitary.

Since $g^{00} \neq 0$, we deduce from (V.2.6) the values of the derivatives $\partial_{00} g_{ij}$ over Σ . For $u^0 \neq 0$, the equation (V.2.7) give the value of $\partial_0 S$ over Σ .

To obtain the values of $\partial_0 p$ and $\partial_0 u^0$ over Σ , we note that (V.2.8) may be written

(V.2.10) $$r \partial_0 u^0 + u^0 r'_p \partial_0 p = \Phi$$

where the value of the right-hand side Φ is known over Σ . The equation (V.2.9) corresponding to $\beta = 0$ may be written

(V.2.11) $$c^2 r f u^0 \partial_0 u^0 - (g^{00} - (u^0)^2) \partial_0 p = 0$$

where the value of ψ is known again over Σ . The equations (V.2.10), (V.2.11) determine the values of $\partial_0 u^0$ and $\partial_0 p$ over Σ if the corresponding determinant is $\neq 0$, that is

(V.2.12) $$(c^2 f r'_p - 1)(u^0)^2 + g^{00} \neq 0$$

If such is the case, $\partial_0 p$, $\partial_0 u^0$ and also the $\partial_0 u^i$ $(i = 1, 2, 3)$ according to (V.2.9) for $\beta = i$ have known values over Σ.

Under the hypothesis

$$g^{00} \neq 0 \qquad u^0 \neq 0 \qquad (c^2 f r'_p - 1)(u^0)^2 + g^0 \neq 0$$

the same conclusions are available for the determination of the consecutive derivatives ; it is sufficient to derive in y^0 the diffuent equations under consideration.

c) We have thus determined the characteristic manifolds of the system (V.1.9), (V.1.10), (V.1.8). If $\varphi = 0$ is the local equation of a regular hypersurface, we obtain three types of characteristic hypersurfaces.

1) The ***gravitational wave fronts,*** solutions of

$$g^{\alpha\beta} \partial_\alpha \varphi \, \partial_\beta \varphi = 0 \tag{V.2.13}$$

2) The hypersurfaces generated by ***the lines of flow,*** solutions of

$$u^\alpha \partial_\alpha \varphi = 0 \tag{V.2.14}$$

hypersurfaces that clearly are time-like. According to (V.2.3), these hypersurfaces are called ***entropy wave fronts***

3) The ***sonic wave fronts,*** solutions of

$$[(\gamma - 1) u^\alpha u^\beta + g^{\alpha\beta}] \, \partial_\alpha \varphi \, \partial_\beta \varphi = 0 \tag{V.2.15}$$

where

(V.2.16) $\gamma = c^2 f r'_p$

According to (V.1.7), the speed v of these fronts is given by

(V.2.17) $v^2 / c^2 = 1/\gamma$

In the following, we suppose that $\tau(p,S)$ *is such that* $\gamma > 1$ that is $v^2 < c^2$. This is equivalent to say that the sonic wave fronts are oriented in the time.

d) It is easy to express the condition $\gamma > 1$ in terms of the function τ (p,S). One has

$$\tau'_p / \tau = f'_p / f - r'_p / r$$

It follows :

$$c^2 \tau'_p = c^2 \mathbf{V} f'_p - c^2 \mathbf{V}^2 f r'_p$$

But $c^2 f'_p = \mathbf{V}$. One has

(V.2.18) $c^2 \tau'_p = - \mathbf{V}^2 (\gamma - 1)$

One has $\gamma > 1$ *if and only if* $\tau'_p < 0$

V.3. - A Leray system and the Cauchy problem.

We will deduce from the system (V.1.8), (V.1.9), (V.1.10) a Leray system.

a) The differential system (V.1.14) of the lines of flow is

(V.3.1) $$c^2 r f u^\alpha \nabla_\alpha u^\beta - (g^{\alpha\beta} - u^\alpha u^\beta)\, \partial_\alpha p = 0$$

Take the contracted covariant derivative ∇_β of (V.3.1).

(V.3.2) $$c^2 rf\; u^\alpha \nabla_\beta \nabla_\alpha u^\beta - (g^{\alpha\beta} - u^\alpha u^\beta)\, \nabla_\alpha \nabla_\beta p =$$

$$= A(1 \text{ in } g_{\lambda\mu}\,,\ 1 \text{ in } p\,,\ 1 \text{ in } S\,,\ 1 \text{ in } u^\lambda)$$

where A depends on the variables and on their derivatives at most at the indicate order. But, according to the Ricci identity

$$\nabla_\beta \nabla_\alpha u^\beta = \nabla_\alpha \nabla_\beta u^\beta + R_{\alpha\rho} u^\rho$$

From the equation (V.2.8) , it follows

$$\nabla_\beta \nabla_\alpha u^\beta = - \nabla_\alpha \left(\frac{r'_p}{r} u^\beta \nabla_\beta p \right) + R_{\alpha\rho} u^\rho$$

and thus

$$u^\alpha \nabla_\beta \nabla_\alpha u^\beta = - \frac{r'_p}{r} u^\alpha u^\beta \; \nabla_\alpha \nabla_\beta p +$$

$$+ B(2 \text{ in } g_{\lambda\mu}\,,\ 1 \text{ in } p\,,\ 1 \text{ in } S\,,\ 1 \text{ in } u^\lambda)$$

According to this relation, (V.3.2) may be written

(V.3.3) $$[(\gamma - 1)\; u^\alpha u^\beta + g^{\alpha\beta}]\, \nabla_\alpha \nabla_\beta p =$$

$$= C(2 \text{ in } g_{\lambda\mu}\,,\ 1 \text{ in } p\,,\ 1 \text{ in } S\,,\ 1 \text{ in } u^\lambda)$$

Now take the derivative of (V.3.3) along the lines of flow by means of the operator $u^\gamma \nabla_j$

$$(V.3.4) \qquad u^\gamma \; [(\gamma - 1) u^\alpha u^\beta + g^{\alpha\beta}] \, \partial_{\alpha\beta\gamma} \; p =$$

$$= E(3 \text{ in } g_{\lambda\mu} , 2 \text{ in } p , 2 \text{ in } S , 2 \text{ in } u^\lambda)$$

We consider now the differential system (V.3.1) of the lines of flow and we apply the differential operator

$$[(\gamma - 1) u^\lambda u^\mu + g^{\lambda\mu}] \nabla_\lambda \nabla_\mu$$

It follows

$$(V.3.5) \qquad c^2 rf u^\alpha [(\gamma - 1) u^\lambda u^\mu + g^{\lambda\mu}] \nabla_\lambda \nabla_\mu \nabla_\alpha u^\beta -$$

$$- (g^{\alpha\beta} - u^\alpha u^\beta) [(\gamma - 1) u^\lambda u^\mu + g^{\lambda\mu}] \nabla_\lambda \nabla_\mu \nabla_\alpha p =$$

$$= F^\beta (3 \text{ in } g_{\rho\sigma} , 2 \text{ in } p , 2 \text{ in } S , 2 \text{ in } u^\rho)$$

From (V.3.3), we deduce

$$(g^{\alpha\beta} - u^\alpha u^\beta) [(\gamma - 1) u^\lambda u^\mu + g^{\lambda\mu}] \, \partial_{\alpha\beta\gamma} \; p =$$

$$= G^\beta (3 \text{ in } g_{\rho\sigma} , 2 \text{ in } p , 2 \text{ in } S , 2 \text{ in } u^\rho)$$

and we obtain

$$(\text{V.3.6}) \qquad c^2 rf\, u^\alpha [(\gamma - 1)\, u^\lambda u^\mu + g^{\lambda\mu}]\, \partial_{\alpha\beta\gamma}\, u^\beta =$$

$$= H^\beta (3 \text{ in } g_{\rho\sigma}, 2 \text{ in } p, 2 \text{ in } S, 2 \text{ in } u^\rho)$$

c) Now, we consider as the unknown functions the $10\ g_{\lambda\mu}$, the scalars p and S' and the $4\ u^\beta$ which define an arbitrary vector. We have thus 16 unknown functions. In a harmonic chart, we have

$$(\text{V.3.7}) \qquad \overset{(h)}{R}_{\alpha\beta} = \chi\, [c^2 rf\, u_\alpha u_\beta - \frac{1}{2} (c^2 rf - 2p)\, g_{\alpha\beta}]$$

and we substitute for these equations their derivatives along the lines of flow. There appears in the right-hand side

$$u^\gamma \nabla_\gamma (c^2 rf)\, u_\alpha u_\beta + c^2 rf\, u^\gamma \nabla_\gamma (u_\alpha u_\beta) - \frac{1}{2} g_{\alpha\beta}\, u^\gamma \nabla_\gamma (rf - 2p) =$$

$$= K_{\alpha\beta} (1 \text{ in } g_{\rho\sigma}, 1 \text{ in } p, 1 \text{ in } S, 1 \text{ in } u^\rho)$$

We consider thus the partial diferential system defined by the derivatives of (V.3.7) along the lines of flow

$$(\text{V.3.8}) \qquad -\frac{1}{2} u^\gamma g^{\alpha\beta}\, \partial_{\alpha\beta\gamma}\, g_{\lambda\mu} = L_{\lambda\mu} \quad (2 \text{ in } g_{\rho\sigma}, 1 \text{ in } p, 1 \text{ in } S, 1 \text{ in } u^\rho)$$

by the equation (V.3.4)

$$(\text{V.3.9}) \qquad u^\gamma [(\gamma - 1)\, u^\alpha u^\beta + g^{\alpha\beta}]\, \partial_{\alpha\beta\gamma}\, p =$$

$$= E (3 \text{ in } g_{\rho\sigma}, 2 \text{ in } p, 2 \text{ in } S, 2 \text{ in } u^\rho)$$

by the equation

(V.3.10) $$u^\gamma \partial_\gamma S = 0$$

and by the system (V.3.6) let

(V.3.11) $$u^\gamma [(\gamma - 1) u^\alpha u^\beta + g^{\alpha\beta}] \partial_{\alpha\beta\gamma} u^\lambda =$$
$$= H^\lambda (3 \text{ in } g_{\rho\sigma}, 2 \text{ in } p, 2 \text{ in } S, 2 \text{ in } u^\rho)$$

This system is a Leray system ; the matrix A , here 16×16 , is the diagonal matrix for which the diagonal elements are respectively

$$a(8) = -\frac{1}{2} u^\gamma g^{\alpha\beta} \partial_{\alpha\beta\gamma}$$

$$a(9) = a(11) = u^\gamma [(\gamma - 1) u^\alpha u^\beta + g^{\alpha\beta}] \partial_{\alpha\beta\gamma}$$

$$a(10) = u^\gamma \partial_\gamma .$$

This system satisfies the hypotheses on the derivatives with the following indices for the unknowns

$s(g_{\alpha\beta}) = 4$ $s(p) = 3$ $s(S) = 3$ $s(u^\gamma) = 3$

and the following indices for the equations

$t(8) = 2$ $t(9) = 1$ $t(10) = 3$ $t(11) = 1$

In fact $a(8)$ et $a(9) = a(11)$ are of order 3 , $a(10)$ is of order 1 . The table giving the maximum order of derivation corresponding to these indices is the following

$$(8)\quad \begin{cases} g_{\alpha\beta} : 2 \\ p : 1 \\ S : 1 \\ u^\gamma : 1 \end{cases} \qquad \begin{matrix}(9)\\(11)\end{matrix} \begin{cases} g_{\alpha\beta} : 3 \\ p : 2 \\ S : 2 \\ u^\gamma : 2 \end{cases} \qquad (10) \begin{cases} g_{\alpha\beta} : 1 \\ p : 0 \\ S : 0 \\ u^\gamma : 0 \end{cases}$$

These maxima orders are compatible with our system

d) *The Cauchy data* $g_{\alpha\beta}$, $\partial_0 g_{\alpha\beta}$, S *corresponding to the system* (V.3.7) , (V.2.7) , (V.2.8) , (V.2.9) *are supposed such that*

(1) *The quadratic form* $g_{\lambda\mu} v^\lambda v^\mu$ *is strictly hyperbolic and* Σ *is spatial at each point for the corresponding elementary cones*

(2) *One has over* Σ

$$F^\rho = 0 \quad \text{for} \quad y^0 = 0$$

(3) *The relations* $S^0_\alpha = \chi\, T^{(f)0}_\alpha$ *and the value of* S *define an admissible value of* p *and thus values for* u^α , *such that* $\gamma > 1$

If C^+_x is the elementary half-cone in the future, its dual is the half-cone Γ^+_x defined by

$$g^{\lambda\mu} \xi_\lambda \xi_\mu \geqslant 0$$

with $\xi(v) \geqslant 0$ if $v \in C^+_x$. To the operator $a(10)$ corresponds the "half-cone" $\Gamma^+_x(10)$ defined by

$$u^\lambda \xi_\lambda \geqslant 0 \qquad\qquad u^\lambda \in C_x^+$$

we consider now the half-cone $\hat{\Gamma}_x^+$ defined by means of

$$[(\gamma - 1)\ u^\lambda u^\mu + g^{\lambda\mu}]\ \xi_\lambda\ \xi_\mu \geqslant 0$$

Under the hypotheses $\gamma > 1$, this half-cone is exterior to the half-cone Γ_x^+ and does not cut the hyperplane $u^\lambda \xi_\lambda = 0$. It follows that the operators a(8) and a(9 $\neq$ a(11) are strictly hyperbolic. We see that the interior of the intersection of the half-cones is Γ_x^+ . Thus the matrix A is strictly hyperbolic.

Suppose that Σ is spatial and consider the system (V.3.7), (V.2.7), (V.2.8), (V.2.9). From the Cauchy data, the equations (V.3.7) give the values of the second derivatives of the $g_{\alpha\beta}$, the equation (V.2.7) the values of the first derivatives of S over Σ . The equations (V.2.8), (V.2.9) give, as we know, the first derivatives of p and of u^α over Σ . By differentiation, (V.3.7) gives the third derivatives of the $g_{\alpha\beta}$, (V.3.3) the second derivatives of p and we deduce from (V.2.9) the values of the second derivatives of u^α . We have obtained the values over Σ of the derivatives of order $\leqslant s(\sigma) - 1$.

We consider the corresponding Cauchy problem for the strictly hyperbolic Leray system (V.3.8) , (V.3.9) , (V.3.10) , (V.3.11). If the initial data are sufficiently regular, the theorem of Leray shows that this Cauchy problem has one and only one solution ($g_{\alpha\beta}$, p , S , u^α) . A standard argument similar to the argument of Section (III.3) give then ***a local existence and uniqueness theorem for the Cauchy problem corresponding to the equations*** (V.1.8) (V.1.9), (V.1.10) ***for a relativistic thermodynamical ideal fluid.***

V.4. - Sonic waves.

One may determine directly the characteristic of the system (V.1.8), (V.1.10) equivalent to the system (V.1.8), (V.1.13), (V.1.14) corresponding to the dynamic of the fluid in consideration. We suppose here that p , S and u^α are of class $\{ C^0$, piecewise C^2). At the traverse of a hypersurface generated by the lines of flow $(u^\alpha \partial_\alpha \varphi = u^\alpha \ell_\alpha = 0)$, the first derivatives of p , S , u^α may be discontinuous. We are led to the following definition : ***a sonic wave is defined by a solution*** (p, S, u^α) ***of the system*** (V.1.8) , (V.1.10) ***of the hydrodynamics such that the derivatives of*** p , S , u^α ***are regularly discontinuous and at least one effectively, at the traverse of a hypersurface*** Σ ***not generated by the lines of flow, which is the wave front.***

a) According to (I.IV.6), we set

$$\text{(V.4.1)} \quad \bar{\delta}[\nabla_\alpha p] = \ell_\alpha \delta p \qquad \bar{\delta}[\nabla_\alpha S] = \ell_\alpha \delta S \qquad \bar{\delta}[\nabla_\alpha u^\beta] = \ell_\alpha \delta u^\beta$$

Study when one at least of the distributions δp , δS , δu^β is not zero

We deduce from (V.1.13) by difference between the two sides of Σ

$$\text{(V.4.2)} \qquad (u^\alpha \ell_\alpha) \, \delta S = 0$$

If $u^\alpha \ell_\alpha = 0$, δS may be $\neq 0$. The relation (V.1.8) may be written explicitely according to (V.1.10) :

$$r \nabla_\alpha u^\alpha + r'_p u^\alpha \partial_\alpha p = 0$$

By difference between the two sides of Σ it follows

(V.4.3) $$r \ell_\alpha \delta u^\alpha + r'_p (u^\alpha \ell_\alpha) \delta p = 0$$

Finally, we deduce from (V.1.14)

(V.4.4) $$c^2 r f (u^\alpha \ell_\alpha) \delta u^\beta - (g^{\alpha\beta} - u^\alpha u^\beta) \ell_\alpha \delta p = 0$$

By multiplication of (V.4.4) by ℓ_β, one obtains according to (V.4.3)

$$\{ c^2 f r'_p (u^\alpha \ell_\alpha)^2 + (g^{\alpha\beta} - u^\alpha \ell^\beta) \ell_\alpha \ell_\beta \} \delta p = 0$$

that is, if $c^2 f r'_p = \gamma$

(V.4.5) $$\{ (\gamma - 1)(u^\alpha \ell_\alpha)^2 + g^{\alpha\beta} \ell_\alpha \ell_\beta \} \delta p = 0$$

b) ***Suppose*** $u^\alpha \ell_\alpha = 0$; δS may be $\neq 0$. According to (V.4.3), (V.4.4), the δu^β may be $\neq 0$ under the condition $\ell_\beta \delta u^\beta = 0$ and $\delta p = 0$. One obtain the so-called ***entropy waves*** for which the fronts are the hypersurfaces generated by the lines of flow.

Suppose $u^\alpha \ell_\alpha \neq 0$. One has $\delta S = 0$. According to (V.4.4), if $\delta p = 0$, one has $\delta u^\beta = 0$ and there is not effective wave. If $\delta p \neq 0$, Σ satisfies the equation of the ***sonic waves***

(V.4.6) $$P(\ell) = (\gamma - 1)(u^\alpha \ell_\alpha)^2 + g^{\alpha\beta} \ell_\alpha \ell_\beta = 0$$

$P(\ell) = 0$ states that if v is the speed of sound, that is the speed of Σ with respect to the fluid, one has

$$v^2/c^2 = 1/\gamma$$

We see that Σ is oriented in time, or that $v < c$, if and only if $\gamma > 1$, which we suppose and which is equivalent to $\tau'_p < 0$.

The fronts of the sonic waves are the hypersurfaces tangent to the cones defined by the tensor :

(V.4.7) $$h^{\alpha\beta} = (\gamma - 1)\ u^\alpha u^\beta + g^{\alpha\beta}$$

For $\gamma > 1$, this tensor defines a quadratic form of strictly hyperbolic type.

V.5. - Main property of the rays in hydrodynamics.

a) The bicharacteristics *-or rays-* associated to the *sonic waves* and which generate the sonic wave fronts are the integral curves of the vector field $N^\beta = h^{\alpha\beta}\ell_\alpha$. These curves are the isotropic geodesics for the metric **h** defined by $h^{\alpha\beta}$. The vector N^β defines the generator of the cone determined by **h** along which the front Σ is tangent to the cone.

One has

(V.5.1) $$N^\beta = \frac{1}{2}\frac{\partial P(\ell)}{\partial \ell_\beta} = h^{\alpha\beta}\ell_\alpha = \ell^\beta + (\gamma-1)(u^\alpha\ell_\alpha)\,u^\beta$$

Let v^β be the component of the velocity of the fluid tangent to Σ. According to $v^\beta\ell_\beta = 0$, one has

(V.5.2) $$u^\beta = v^\beta + ((u^\alpha\ell_\alpha)/(\ell^\alpha\ell_\alpha)\,\ell^\beta$$

By substitution in (V.5.1), we obtain

(V.5.3) $$N^{\beta} = (\gamma - 1)\,(u^{\alpha} \ell_{\alpha})\, v^{\beta}$$

and the direction of N^{β} is naturally the direction of v^{β}.

b) We adopt the hypotheses of Section (I.6). The metric is supposed of class C^2 ; p, S, u^{λ} are continuous over the domain Ω and of class C^2 over Ω^+ and Ω^- and their first and second derivatives are regularly discontinuous at the traverse of Σ.

We will show that δp (and also the δu^{λ}) are propagating along the rays, that is δp satisfies a differential system of the form

$$N^{\beta} \nabla_{\beta} \delta p + A \delta p = 0$$

From (I.6.6) it follows that there exist distributions $\overline{p}$, $\overline{S}$, $\overline{u}^{\lambda}$ with support over Σ such that

(V.5.4) $$\overline{\delta}[\nabla_{\alpha} \nabla_{\beta} p] = \nabla_{\alpha} \ell_{\beta}\, \delta p + \ell_{\alpha} \nabla_{\beta} \delta p + \ell_{\beta} \nabla_{\alpha} \delta p + \ell_{\alpha} \ell_{\beta}\, \overline{p}$$

(V.5.6) $$\overline{\delta}[\nabla_{\alpha} \nabla_{\beta} S] = \ell_{\alpha} \ell_{\beta}\, \overline{S} \qquad (\delta S = 0)$$

(V.5.7) $$\overline{\delta}[\nabla_{\alpha} \nabla_{\beta} u^{\lambda}] = \nabla_{\alpha} \ell_{\beta}\, \delta u^{\lambda} + \ell_{\alpha} \nabla_{\beta} \delta u^{\lambda} + \ell_{\beta} \nabla_{\alpha} \delta u^{\lambda} + \ell_{\alpha} \ell_{\beta}\, \overline{u}^{\lambda}$$

(V.1.8) may be written

(V.5.8) $$c^2 r f\, \nabla_{\alpha} u^{\alpha} + \gamma\, u^{\alpha} \partial_{\alpha} p = 0$$

Dividing by c^2rf and differentiating, it follows

$$\nabla_\beta \nabla_\alpha u^\alpha + (\gamma/c^2rf)\, u^\alpha \nabla_\beta \nabla_\alpha p + \nabla_\beta ((\gamma/c^2rf)\, u^\alpha) \nabla_\alpha p = 0$$

By difference between the two sides of Σ, one obtains

$$\bar{\delta}[\nabla_\beta \nabla_\alpha u^\alpha] + (\gamma/c^2rf) u^\alpha \bar{\delta}[\nabla_\beta \nabla_\alpha p] + \bar{\delta}[\nabla_\beta ((\gamma/c^2rf)\, u^\alpha) \nabla_\alpha p] = 0$$

But

$$\bar{\delta}[\nabla_\beta ((\gamma/c^2rf)\, u^\alpha) \nabla_\alpha p] =$$

$$= (\nabla_\alpha p)_+ \bar{\delta}[\nabla_\beta ((\gamma/c^2rf)\, u^\alpha)] + \{\nabla_\beta ((\gamma/c^2rf)\, u^\alpha\}_- \bar{\delta}[\nabla_\alpha p]$$

is a linear combination of terms in δp and δu^λ and is thus proportional to δp. One may write

$$\bar{\delta}[\nabla_\beta \nabla_\alpha u^\alpha] + (\gamma/c^2rf)\, u^\alpha \bar{\delta} \nabla_\beta \nabla_\alpha p \simeq 0 \qquad \text{(V.5.9)}$$

where $\simeq$ means ***modulo terms proportional to*** δp.

Similarly, for the entropy, one has

$$u^\alpha \bar{\delta}[\nabla_\alpha \nabla_\beta S] \simeq 0$$

that is according to (V.5.6)

$$u^\alpha \ell_\alpha \ell_\beta \bar{S} \simeq 0$$

Therefore $\overline{S}$ is $\simeq$ 0 and it follows :

(V.5.10) $$\overline{\delta}\,[\nabla_\alpha \nabla_\beta S\,] \simeq 0$$

Finally, with the same argument over (V.1.14), it follows

(V.5.11) $$c^2 rf\; u^\alpha\, \overline{\delta}\; [\,\nabla_\beta \nabla_\alpha u^\beta\,] - (g^{\alpha\beta} - u^\alpha u^\beta)\; \overline{\delta}\,[\nabla_\alpha \nabla_\beta p\,] \simeq 0$$

According to (V.5.9) one obtains :

$$\gamma\, u^\alpha u^\beta\, \overline{\delta}\,[\nabla_\alpha \nabla_\beta p\,] + (g^{\alpha\beta} - u^\alpha u^\beta\;)\, \overline{\delta}\,[\nabla_\alpha \nabla_\beta p\,] \simeq 0$$

From (V.5.4) it follows

$$h^{\alpha\beta}\,(\ell_\alpha \nabla_\beta \delta p + \ell_\beta \nabla_\alpha \delta p + \ell_\alpha \ell_\beta \overline{p}\,) \simeq 0$$

But Σ being the front of a sonic wave, one has $h^{\alpha\beta}\ell_\alpha \ell_\beta = 0$ and one obtains

$$N^\beta \nabla_\beta \delta p \simeq 0$$

We have the following theorem that expresses the main property of the rays.

Theorem. *Under the hypotheses of Section* (I.6)c , *the infinitesimal discontinuities* δp , δu^λ *corresponding to a sonic wave are propagating along the associate rays according to the differential system*

$$N^{\beta}\,\nabla_{\beta}\,\delta p \simeq 0 \qquad\qquad N^{\beta}\,\nabla_{\beta}\,\delta u^{\lambda} \simeq 0$$

c) Under the same hypotheses, we will show some invariance properties in the infinitesimal shocks given by the sonic waves. We have seen that the operator δ is a derivation. The metric and φ being supposed respectively of class C^1 and C^2 over Ω, one has

$$\delta g^{\alpha\beta} = 0 \qquad\qquad \delta \ell_{\alpha} = 0$$

By difference of (V.1.8) between the two sides of Σ, it follows

(V.5.12) $$\delta(r u^{\alpha} \ell_{\alpha}) = 0$$

Therefore *the scalar* $a = r u^{\alpha} \ell_{\alpha}$ *is invariant under* δ.

In the same way, by difference of (V.1.10) between the two sides of Σ, one obtains

$$\delta(T^{(f)\,\alpha\beta}\,\ell_{\alpha} = 0$$

and the vector

(V.5.13) $$W^{\beta} = c^2 r f\,(u^{\alpha}\ell_{\alpha})\;u^{\beta} - p\,\ell^{\beta} = c^2\,a\,f\,u^{\beta} - p\ell^{\beta}$$

is invariant under δ. This vector may be written

(V.5.14) $$W^{\beta} = c^2\,a\,f\,v^{\beta} + (c^2\,a\,f\,(u^{\alpha}\ell_{\alpha}\,/\,\ell^{\alpha}\ell_{\alpha}) - p\,)\,\ell^{\beta}$$

where v^{β} is the tangential component of u^{β} (see (V.5.2)) ; $W^{\beta}\,\ell_{\beta}$ is invariant under δ, that is the scalar

$$c^2 a f (u^\alpha \ell_\alpha / \ell^\alpha \ell_\alpha) - p$$

and one has

(V.5.15) $\delta(f v^\beta) = 0$

In particular ***the direction of the ray given by the direction of*** v^β ***is invariant under the operator*** δ ***of infinitesimal discontinuity.***

V.6. - Thermoynamic hypotheses.

a) Many years ago, I have adopted, for the relativistic thermodynamic ideal fluids, the following hypotheses on the state function $\tau(p, S)$. In the considered domain of the variables p and S we suppose

(H_1) *One has* $\tau'_p < 0$ *and* $\tau''_{p^2} > 0$

(H_2) τ'_S *is* $\neq 0$ *and there exists a function* $S = S(p, \tau)$ *whose unique inverse is* $\tau = \tau(p, S)$.

We know that $\tau'_p < 0$ is equivalent to $v < c$ or to the fact that the sonic wave fronts are oriented in the time ; $\tau''_{p^2} > 0$ is a convexity condition related to the stability of hydrodynamic shock waves with respect to sonic waves ; (H_2) expresses the independance of the thermodynamical variables in consideration of the fluid.

The hypotheses (H_1) et (H_2) correspond for the classical approximation at the compressibility hypotheses introduced by Hermann

Weyl in the theory of the classical hydrodynamic shock waves, hypotheses that are the following : $V'_p < 0$, $V''_{p^2} > 0$ and $V'_s > 0$.

Israel [5] and Lucquiaud [13] have established by different methods that for a relativistic Boltzmann gaz, one necessarily has $\tau'_p < 0$, $\tau''_{p^2} > 0$ and $\tau'_s > 0$. We shall also be interested in the case $\tau'_s < 0$ which may be valid for condensed matter and this case may be present in astrophysics.

b) Inversing the function $\tau = \tau\,(p\,,\,S)$, one obtains a function $p = p\,(\tau\,,\,S)$ expressing the pressure in terms of the variables τ and S . One identically has

(V.6.1) $$p = p\ \{\,\tau\,(p\,,S)\,,\,S\,\}$$

Study the partial derivatives of $p\,(\tau\,,\,S)$. By differentiation of (V.6.1) with respect to p , with a constant S , one has

(V.6.2) $$p'_\tau\,\tau'_p = 1$$

It follows

(V.6.3) $$p'_\tau = \frac{1}{\tau'_p}$$

and $p'_\tau < 0$.

By differentiation of (V.6.2) with respect to p , with a constant S , one has

$$p''_{\tau^2}\,(\tau'_p)^2 + p'_\tau\,\tau''_{p^2} = 0$$

It follows

(V.6.4) $$p''_{\tau^2} = -\frac{\tau''_{p^2}}{(\tau'_p)^3}$$

and thus $p''_{\tau^2} > 0$. In a similar way, one has

(V.6.5) $$p'_S = -\frac{\tau''_S}{\tau'_p}$$

c) We shall not study hydrodynamic shock waves in themselves, but we consider their theory as a special case of the complete study of magnetohydrodynamic shock waves (see Chapter VII).

VI - THE EQUATIONS OF MAGNETOHYDRODYNAMICS.

VI.1 - Electromagnetic field with induction.

a) Inside matter, a general electromagnetic field is determined by two 2-forms H and G, that is to say by two skew symmetric tensor fields of order 2. The tensor H is called here, ***the electric field-magnetic induction tensor*** and the tensor G ***the electric induction-magnetic field tensor.***

Let $*H$ and $*G$ be the dual 2-forms of H and G in the space-time (V_4, g), where we have chosen ***an orientation.*** If η is the volume element 4-form of (V_4, g) corresponding to the orientation, we have

$$(\text{VI.1.1}) \qquad (*H)_{\alpha\beta} = \frac{1}{2}\,\eta_{\lambda\mu\alpha\beta}\,H^{\lambda\mu} \qquad\qquad (*G)_{\alpha\beta} = \frac{1}{2}\,\eta_{\lambda\mu\alpha\beta}\,G^{\lambda\mu}$$

We note that with our signature of **g**, we have for the 2-forms $*^2 = -1$. If **u** is a unitary vector of V_4 defining a temporal direction at $x \in V_4$, we set

$$(\text{VI.1.2}) \qquad \left\{ \begin{array}{l} e_\beta = u^\alpha H_{\alpha\beta} \\ \\ b_\beta = u^\alpha (*H)_{\alpha\beta} \end{array} \right. \qquad \left\{ \begin{array}{l} d_\beta = u^\alpha G_{\alpha\beta} \\ \\ h_\beta = u^\alpha (*G)_{\alpha\beta} \end{array} \right.$$

These four vectors are orthogonal to **u**. The vectors **e** and **d** are called respectively ***the electric field and the electric induction*** corresponding to the unitary vector **u**. The vectors **h** and **b** are called respectively ***the magnetic field and the magnetic induction*** corresponding to **u** . We will see that **e** and **b** determine H and that **d** and **h** determine G .

b) Consider at $x \in \mathbf{V}_4$ an orthonormalized frame $\{v_{(\rho)}\}$ such that $v_{(0)} = \mathbf{u}$; $\{v_{(\rho)}\}$ is a rest frame with respect to **u** . In this frame, the components e_0 , d_0 , b_0 , h_0 of our four vectors are zero and we have $(i, j, k = 1, 2, 3)$

$$\text{(VI.1.3)} \qquad \begin{cases} e_i = H_{0i} \\ b_i = (* H)_{0i} \end{cases} \qquad \begin{cases} d_i = G_{0i} \\ h_i = (* G)_{0i} \end{cases} \qquad (i, j, k, \ldots = 1, 2, 3)$$

It follows that :

$$\text{(VI.1.4)} \qquad \begin{cases} e^i = H^{0i} \\ b^i = (* H)^{0i} \end{cases} \qquad \begin{cases} d^i = G^{0i} \\ h^i = (* G)^{0i} \end{cases}$$

But, according to the definition of the operator $*$, one has

$$\begin{cases} (* H)_{0i} = \dfrac{1}{2}\, \eta_{0ijk}\, H^{jk} \\ \\ H_{0i} = -\dfrac{1}{2}\, \eta_{0ijk}\, (*H)^{jk} \end{cases} \qquad \begin{cases} (* G)_{0i} = \dfrac{1}{2}\, \eta_{0ijk}\, G^{jk} \\ \\ G_{0i} = -\dfrac{1}{2}\, \eta_{0ijk}\, (*G)^{jk} \end{cases}$$

We deduce from (VI.1.3) the following relations :

$$\text{(VI.1.5)} \qquad \begin{cases} e_i = -(* H)_{jk} \\ \\ b_i = H_{jk} \end{cases} \qquad \begin{cases} d_i = -(* G)_{jk} \\ \\ h_i = G_{jk} \end{cases}$$

where the permutation (i, j, k) of $(1, 2, 3)$ is supposed be even. Similarly one has

(VI.1.6)
$$\begin{cases} e^i = (*H)^{jk} \\ b^i = -H^{jk} \end{cases} \qquad \begin{cases} d^i = (*G)^{jk} \\ h^i = -G^{jk} \end{cases}$$

It is now easy to express the two tensors H and G in terms of the four vectors $\mathbf{e}$, $\mathbf{d}$, $\mathbf{b}$, $\mathbf{h}$. In the rest frame $\{v_{(\rho)}\}$ where $u^0 = 1$, $u^i = 0$, we may rewrite $H_{0i} = e_i$ as

$$H_{0i} = u_0\, e_i - u_i\, e_0 - \eta_{0ijk}\, u^j\, b^k$$

and we may also rewrite $H_{jk} = -\eta_{0ijk}\,(*H)^{0i} = -\eta_{0ijk}\, b^i$ as

$$H_{jk} = u_j\, e_k - u_k\, e_j - \eta_{\lambda\mu jk}\, u^\lambda\, b^\mu$$

One thus obtain the unique tensor formula

(VI.1.7)
$$H_{\alpha\beta} = u_\alpha\, e_\beta - u_\beta\, e_\alpha - \eta_{\lambda\mu\alpha\beta} u^\lambda\, b^\mu$$

Similarly one has

(VI.1.8)
$$G_{\alpha\beta} = u_\alpha\, d_\beta - u_\beta\, d_\alpha - \eta_{\lambda\mu\alpha\beta} u^\lambda\, h^\mu$$

We may write (VI.1.7), (VI.1.8) in terms of exterior products

(VI.1.9)
$$H = u \wedge e - *(u \wedge b) \qquad G = u \wedge d - *(u \wedge h)$$

and by the action of the operator $*$, one obtains

(VI.1.10) $*H = u \wedge b - *(u \wedge e) \qquad *G = u \wedge h - *(u \wedge d)$

c) In the theory of Maxwell, the inductions depend linearly on the fields and are given by ***constitutive equations ;*** these equations depend on the nature of the matter. We consider here the ***isotropic case*** and we write as constitutive equations

(VI.1.11) $d = \lambda e \qquad\qquad b = \mu h$

λ is here the ***dielectric permitivity*** of the matter and μ its ***magnetic permeability.*** From these relations, it is possible to deduce a simple expression for G in terms of H. According to (VI.1.9), we have [8]:

$$G = \lambda(u \wedge e) - \frac{1}{\mu} * (u \wedge b)$$

But

$$- * (u \wedge b) = H - u \wedge e$$

It follows that

(VI.1.12) $$G = \frac{1}{\mu} H + \frac{\lambda\mu - 1}{\mu} u \wedge e$$

or more explicitly

(VI.1.13) $$G_{\alpha\beta} = \frac{1}{\mu} H_{\alpha\beta} + \frac{\lambda\mu - 1}{\mu} (u_\alpha u^\rho H_{\rho\beta} - u_\beta u^\rho H_{\rho\alpha})$$

VI.2. - Maxwell equations and energy tensor of Minkowski.

a) We suppose that, in the considered dmain of V_4, the electromagnetic field satisfies Maxwell's equations :

(VI.2.1) $$dH \quad \text{or} \quad \varepsilon(*H) = 0$$

and

(VI.2.2) $$\varepsilon G = J$$

where J is the electric current ; (VI.2.2) implies

$$\varepsilon J = 0$$

The electric current is the sum of two terms corresponding respectively to the convection current and to the conduction current

(VI.2.3) $$J^{\beta} = \varepsilon u^{\beta} + \sigma\, u_{\alpha}\, H^{\alpha\beta} = \varepsilon\, u^{\beta} + \sigma\, e^{\beta}$$

where ε is the ***proper density of electric charge*** and σ the ***conductivity*** of the fluid.

b) Let us consider Minkowski's energy tensor, defined by [8]

(VI.2.4) $$\tau_{\alpha\beta} = \frac{1}{4}\, g_{\alpha\beta}\, H_{\rho\sigma}\, G^{\rho\sigma} - H_{\alpha\rho}\, G_{\rho}{}^{\rho}$$

This tensor is in general not symmetrical. The non symetry of Minkowski's tensor has given long discussions in the litterature. Other tensors have been considered. But we will see that the tensor of Minkowski is symmetrical in the case of the relativistic magnetohydrodynamics and the other tensors in consideration are not different in this case from the tensor of Minkowski. We shall adopt this energy tensor here, without discussion.

VI.3. - The case of the magnetohydrodynamics.

a) Magnetohydrodynamics (MHD) is here the study of the properties of an ideal fluid of ***infiite conductivity*** $\sigma = \infty$. The electric current J and thus the product σe being essentially finite, one has necessaryly i this case $e = 0$. ***The electromagnetic field is reduced with respect to the fluid, to its magnetic part.*** We suppose that the fluid admits ***a constant magnetic permeability*** μ.

It follows from (VI.1.12) that

$$G_{\alpha\beta} = \frac{1}{\mu} H_{\alpha\beta} \tag{VI.3.1}$$

and from (VI.1)

$$(*H)_{\alpha\beta} = \mu \left(u_\alpha h_\beta - u_\beta h_\alpha \right) \tag{VI.3.2}$$

b) Evaluate $\tau_{\alpha\beta}$ given by (VI.2.4). Oe has

$$H_{\alpha\rho} G^{\beta\rho} = \eta_{\alpha\rho\lambda\mu} u^\lambda b^\mu \eta^{\beta\rho\nu\sigma} u_\nu h_\sigma$$

let

$$H_{\alpha\rho} G^{\beta\rho} = - \varepsilon^{\beta\nu\sigma}_{\alpha\lambda\mu} u^\lambda b^\mu u_\nu h_\sigma$$

where ε is the tensor indicator of Kroneker. For $\alpha = \beta$, one obtains the term

$$- g_\alpha^\beta\, b^\nu\, h_\nu \;;$$

for $\alpha = \nu$, one has

$$u_\alpha\, u^\beta\, b^\nu\, h_\nu$$

and for $\alpha = \sigma$

$$h_\alpha\, b^\beta$$

One obtains :

$$H_{\alpha\rho}\, G^{\beta\rho} = - g_\alpha^\beta\, b^\nu\, h_\nu + u_\alpha\, u^\beta\, b^\nu\, h_\nu + h_\alpha\, b^\beta$$

and after contraction between α and β , one has

$$H_{\rho\sigma}\, G^{\rho\sigma} = - 4\, b^\nu\, h_\nu + 2\, b^\nu\, h_\nu = -2\, b^\nu\, h_\nu$$

It follows :

$$\tau_{\alpha\beta} = - \frac{1}{2}\, \mu\, (h^\nu\, h_\nu)\, g_{\alpha\beta} - \mu (h^\nu\, h_\nu)\, (u_\alpha\, u_\beta - g_{\alpha\beta}) - \mu\, h_\alpha\, h_\beta$$

h being a spatial vector orthogonal to u , we set :

(VI.3.3) $$|\, h\,|^2 = - h^\nu\, h_\nu > 0$$

and we obtain

(VI.3.4) $$\tau_{\alpha\beta} = \mu\, |\, h\,|^2\, u_\alpha\, u_\beta - \frac{1}{2}\, \mu\, |\, h\,|^2\, g_{\alpha\beta} - \mu\, h_\alpha\, h_\beta$$

The tensor $\tau_{\alpha\beta}$ should be symmetrical, since H is proportional to G , and we verify over (VI.3.4) that such is the case.

VI.4. - The main equations of relativistic magneto-hydrodynamics.

a) In the following, we shall consider a relativistic ideal fluid with a magnetic permeability μ = const. and an infinite conductivity σ in an electromagnetic field. The notations coincide with the notations of Chap. V. The total energy tensor is here the sum of the energy tensor of the fluid and of the energy tensor of the electromagnetic field. One has [9]

$$T_{\alpha\beta} = T^{(f)}_{\alpha\beta} + \tau_{\alpha\beta}$$

where $T^{(f)}_{\alpha\beta}$ is given by (V.1.4) and $\tau_{\alpha\beta}$ by (VI.3.4). One has

$$T_{\alpha\beta} = (c^2 rf + \mu |h|^2)\, u_\alpha u_\beta - (p + \frac{1}{2} \mu |h|^2)\, g_{\alpha\beta} - \mu h_\alpha h_\beta \tag{VI.4.1}$$

It follows

$$T_{\alpha\beta} u^\beta = (c^2 rf - p + \frac{1}{2} \mu |h|^2)\, u_\alpha$$

The proper energy density of our system (fluid + field) is (compared to Sect. (V.1))

$$c^2 rf - p + \frac{1}{2} \mu |h|^2 = \rho + \frac{1}{2} \mu |h|^2$$

where ρ is the dynamic part and $(1/2)\, \mu |h|^2$ the magnetic part.

We may interest the coefficient $q = p + \frac{1}{2} \mu\, h^2|$ of $g_{\alpha\beta}$ in (VI.4.1) saying that the magnetic field gives here a supplementary pressure equal to $(1/2)\,\mu\,|h|^2$.

b) In the case of the magnetohydrodynamics, the electric current J is not known, but it is only defined by εG. The Maxwell equations are reduced to (VI.2.1) thus is, according to (VI.3.2), to

(VI.4.2) $$\nabla_\alpha (u^\alpha h^\beta - u^\beta h^\alpha) = 0$$

The differential system of relativistic M H D is given by the following equations : first, ***the equation of conservation of matter***

(VI.4.3) $$\nabla_\alpha (r\, u^\alpha) = 0$$

then the ***Maxwell equations*** reduced to (VI.4.2) ***and finally the equations of the relativistic dynamics***

(VI.4.4) $$\nabla_\alpha T^{\alpha\beta} = 0$$

where $T_{\alpha\beta}$ is given by (VI.4.1).

In general relativisty (VI.4.4) is a consequence of the Einstein equations. In making the present abstraction from Einstein equations, we are led to suppose that the space-time is given

c) The Maxwell equations (VI.4.2) may be written

(VI.4.5) $$h^\beta \nabla_\alpha u^\alpha + u^\alpha \nabla_\alpha h^\beta - h^\alpha \nabla_\alpha u^\beta - u^\beta \nabla_\alpha h^\alpha = 0$$

From these equations, one may deduce two interesting consequences through the scalar product with either u_β or with h_β. According to $h^\beta u_\beta = 0$ and the unitary character of u_β, one obtains by product of (VI.4.5) by u_β:

$$\text{(VI.4.6)} \qquad u^\alpha u^\beta \nabla_\alpha h_\beta - \nabla_\alpha h^\alpha = 0$$

Similarly one obtains by product of (VI.4.5) by h_β (according to $|h|^2 = - h^\beta h_\alpha$)

$$\text{(VI.4.7)} \qquad \frac{1}{2} u^\alpha \nabla_\alpha |h|^2 + |h|^2 \nabla_\alpha u^\alpha - h^\alpha u^\beta \nabla_\alpha h_\beta = 0$$

VI.5. - The equation of continuity and the system of the lines of flow.

a) According to (VI.4.1), the system (VI.4.4) may be written

$$\text{(VI.5.1)} \qquad \nabla_\alpha \{(c^2 rf + \mu |h|^2) u^\alpha\} u_\beta + (c^2 rf + \mu |h|^2) u^\alpha \nabla_\alpha u_\beta -$$

$$- \partial_\beta (p + \frac{1}{2} \mu h^2) - \mu \nabla_\alpha h^\alpha h_\beta - \mu h^\alpha \nabla_\alpha h_\beta = 0$$

Multiplying by u^β, it follows

$$\nabla_\alpha \{ (c^2 rf + \mu |h|^2) u^\alpha \} - u^\alpha \partial_\alpha (p + \frac{1}{2} \mu |h|^2) - \mu h^\alpha u^\beta \nabla_\alpha h_\beta - 0$$

that is to say

$$\nabla_\alpha (c^2 rf\ u^\alpha) - u^\alpha \partial_\alpha p + \mu \{\frac{1}{2} u^\alpha \nabla_\alpha |h|^2 + |h|^2 \nabla_\alpha u^\alpha - h^\alpha u^\beta \nabla_\alpha h_\beta \} = 0$$

that may be written, according to (VI.4.7) :

$$\nabla_\alpha (c^2 r f u^\alpha) - u^\alpha \partial_\alpha p = 0 \tag{VI.5.2}$$

relation identical with (V.1.12). This equation implies always (see Section (V.I)) the ***equation of adiabatic flow*** [9]

$$u^\alpha \partial_\alpha S = 0 \tag{VI.5.3}$$

b) The equations (VI.5.1) may be then written

$$(c^2 r f + \mu |h|^2) u^\alpha \nabla_\alpha u^\beta - (g^{\alpha\beta} - u^\alpha u^\beta) \partial_\alpha (p + \frac{1}{2} \mu |h|^2) \tag{VI.5.4}$$

$$+ \mu h^\lambda u^\mu \nabla_\lambda h_\mu u^\beta - \mu \nabla_\alpha h^\alpha h^\beta - \mu h^\alpha \nabla_\alpha h^\beta = 0$$

which are ***the equations of the lines of flow.*** These equations may be written according to (VI.4.7)

$$(c^2 r f + \mu |h|^2) u^\alpha \nabla_\alpha u^\beta - (g^{\alpha\beta} - u^\alpha u^\beta) \partial_\alpha p - \frac{1}{2} \mu g^{\alpha\beta} \nabla_\alpha |h|^2 + \tag{VI.5.5}$$

$$+ \mu u^\alpha u^\beta \nabla_\alpha |h|^2 + \mu |h|^2 \nabla_\alpha u^\alpha u^\beta - \mu \nabla_\alpha h^\alpha h^\beta - \mu h^\alpha \nabla_\alpha h^\beta = 0$$

Multiplying by h_β , we obtain

$$(c^2 r f + \mu |h|^2) u^\alpha h^\beta \nabla_\alpha u_\beta - h^\alpha \partial_\alpha p - \frac{1}{2} \mu h^\alpha \nabla_\alpha |h|^2 + \mu |h|^2 \nabla_\alpha h^\alpha$$

$$+ \frac{1}{2} \mu h^\alpha \nabla_\alpha |h|^2 = 0$$

let

$$(c^2 rf + \mu |h|^2)\, u^\alpha h^\beta \nabla_\alpha h_\beta + h^\alpha \partial_\alpha p + \mu |h|^2 \nabla_\alpha h^\alpha = 0$$

According to (VI.4.6), it follows

(VI.5.6) $$c^2 rf \nabla_\alpha h^\alpha + h^\alpha \partial_\alpha p = 0$$

The fundamental system (VI.4.2), (VI.4.3), (VI.4.4) of magnetohydrodynamics is equivalent to the system formed by (VI.4.3), (VI.5.3), (VI.5.4) and the explicit Maxwell equations (VI.4.5).

VI.6. - Magnetosonic waves and Alfven waves.

a) Suppose that $(p, S, u^\alpha, h^\alpha)$ are of class (C^0, piecewise C^1) on a domain Ω. We study a hypersurface Σ of Ω such that the derivatives of p, S, u^α, h^α may be regularly discontinuous and at least one effectively, at the traverse of Σ. There exist distributions δp, δS, δu^α, δh^α with support over Σ such that ($\ell = d\varphi$)

$$\begin{cases} \bar{\delta}[\nabla_\alpha, p] = \ell_\alpha \delta p \\ \bar{\delta}[\nabla_\alpha, u^\beta] = \ell_\alpha \delta u^\beta \end{cases} \qquad \begin{cases} \bar{\delta}[\nabla_\alpha, S] = \ell_\alpha \delta S \\ \bar{\delta}[\nabla_\alpha, h^\beta] = \ell_\alpha \delta h^\beta \end{cases}$$

The relations (VI.4.3) and (VI.5.3) give exactly as in the hydrodynamic case

(VI.6.1) $$c^2 rf\, \delta(u^\alpha \ell_\alpha) + \gamma(u^\alpha \ell_\alpha)\, \delta p = 0$$

and

(VI.6.2) $(u^\alpha \ell_\alpha)\,\delta S = 0$

One deduces from the Maxwell equations (VI.4.5)

(VI.6.3) $u^\beta\,\delta(h^\alpha \ell_\alpha) + (h^\alpha \ell_\alpha)\,\delta u^\beta - (u^\alpha \ell_\alpha)\,\delta h^\beta - h^\beta \delta(u^\alpha \ell_\alpha) = 0$

The differential system of the lines of flow gives

$$(c^2 rf + \mu |h|^2)(u^\alpha \ell_\alpha)\,\delta u^\beta - (g^{\alpha\beta} - u^\alpha u^\beta)\,\ell_\alpha\,\delta p - \frac{1}{2}\mu \ell^\beta \delta |h|^2$$

(VI.6.4) $$+ \mu(u^\alpha \ell_\alpha)\,u^\beta \delta |h|^2 + \mu |h|^2 \delta(u^\alpha \ell_\alpha)\,u^\beta$$

$$- \mu\,\delta(h^\alpha \ell_\alpha)\,h^\beta - \mu(h^\alpha \ell_\alpha)\,\delta h^\beta = 0$$

and (VI.5.5) gives :

(VI.6.5) $c^2 rf\,\delta(h^\alpha \ell_\alpha) + (h^\alpha \ell_\alpha)\,\delta p = 0$

b) ***Suppose*** $u^\alpha \ell_\alpha = 0$; δS may be $\neq 0$; (VI.6.1) gives $\delta(u^\alpha \ell_\alpha) = 0$. The formulas (VI.6.3), (VI.6.4), (VI.6.5) are reduced to

(VI.6.6) $u^\beta\,\delta(h^\alpha \ell_\alpha) + (h^\alpha \ell_\alpha)\,\delta u^\beta = 0$

(VI.6.7) $-\ell^\beta \delta\left(p + \frac{1}{2}\mu |h|^2\right) - \mu\,\delta(h^\alpha \ell_\alpha)\,h^\beta - \mu(h^\alpha \ell_\alpha)\,\delta h^\beta = 0$

(VI.6.8) $c^2 rf\,\delta(h^\alpha \ell_\alpha) + (h^\alpha \ell_\alpha)\,\delta p = 0$

Multiplying (VI.6.7) by h_β , it follows

(VI.6.9) $$- h^\alpha \ell_\alpha \delta p + \mu |h|^2 \delta (h^\alpha \ell_\alpha) = 0$$

We deduce from (VI.6.8), (VI.6.9) $\delta p = 0$, $\delta(h^\alpha \ell_\alpha) = 0$ and $\delta |h|^2 = 0$

c) ***For*** $u^\alpha \ell_\alpha$ ***arbitrary,*** Maxwell's equations imply :

$$(u^\alpha \ell_\alpha)\, u^\beta \delta h_\beta - \delta(h^\alpha \ell_\alpha) = 0$$

$$\frac{1}{2}(u^\alpha \ell_\alpha)\, \delta |h|^2 + |h|^2 \delta(u^\alpha \ell_\alpha) - (h^\alpha \ell_\alpha)\, u^\beta \delta h_\beta = 0$$

which by elimination of $u^\beta \delta h_\beta$ gives

(VI.6.10) $$\frac{1}{2}(u^\alpha \ell_\alpha)^2 \ \delta |h|^2 + |h|^2 (u^\alpha \ell_\alpha)\, \delta(u^\alpha \ell_\alpha) - (h^\alpha \ell_\alpha)\, \delta(h^\alpha \ell_\alpha) = 0$$

Finally, multplying (VI.6.4) by ℓ_β and taking the preceding relations in account, one obtains

(VI.6.11) $$c^2 r f (u^\alpha \ell_\alpha)\, \delta(u^\alpha \ell_\alpha) - (g^{\alpha\beta} - u^\alpha u^\beta)\, \ell_\alpha \ell_\beta \delta p - \\ - \frac{1}{2} \mu \ell^\beta \ell_\beta \delta |h|^2 = 0$$

Eliminating $\delta |h|^2$ between (VI.6.10) and (VI.6.11), there results

(VI.6.12) $$[c^2 r f (u^\alpha \ell_\alpha)^2 + \mu |h|^2 \ell^\beta \ell_\beta]\, (u^\alpha \ell_\alpha)\, \delta(u^\alpha \ell_\alpha) - \\ - \mu (\ell^\beta \ell_\beta)(h^\alpha \ell_\alpha)\, \delta(h^\alpha \ell_\alpha)(u^\alpha \ell_\alpha)^2 (\ell^\beta \ell_\beta - (u^\beta \ell_\beta)^2)\, \delta p = 0$$

The three linear equations with the unknowns $\delta(u^\alpha \ell_\alpha)$, δp, $\delta(h^\alpha \ell_\alpha)$ given by (VI.6.12), (VI.6.1) and (V.6.5) imply that these unknowns are zero if the corresponding determinant Δ is $\neq 0$.

This determinant may be writen

$$\Delta / c^2 rf = \{ c^2 rf (u^\alpha \ell_\alpha)^2 + \mu |h|^2 \ell^\beta \ell_\beta \} (u^\alpha \ell_\alpha)^2 \gamma +$$

$$+ c^2 rf (u^\alpha \ell_\alpha)^2 (\ell^\beta \ell_\beta - (u^\alpha \ell_\alpha)^2) - \mu (h^\alpha \ell_\alpha)^2 \ell^\beta \ell_\beta$$

Si $\Delta = 0$, δp may be $\neq 0$; the relations (VI.6.1), (VI.6.5) give then $\delta(u^\alpha \ell_\alpha)$ and $\delta(h^\alpha \ell_\beta)$ and (VI.6.10) gives $\delta |h|^2$.

d) Our study has been relative to p and to the transverse components of u^β and h^β with respect to Σ. Now decompose u^β and h^β in tangential and transverse components. One has

$$u^\beta = v^\beta + (u^\alpha \ell_\alpha / \ell^\alpha \ell_\alpha) \ell_\beta \qquad h^\beta = k^\beta + (k^\alpha \ell_\alpha)/(\ell^\alpha \ell_\alpha) \ell^\beta$$

where $v^\beta \ell_\beta = 0$, $h^\beta \ell_\beta = 0$. From (VI.6.3) and (VI.6.4) there results :

(VI.6.13) $$(h^\alpha \ell_\alpha) \delta v^\beta - (u^\alpha \ell_\alpha) \delta k^\beta \simeq 0$$

(VI.6.14) $$(c^2 rf + \mu |h|^2) (u^\alpha \ell_\alpha) \delta v^\beta - \mu (h^\alpha \ell_\alpha) \delta k^\beta \simeq 0$$

where $\simeq$ signifies modulo terms in δp. The determinant of the equations (VI.6.13). (VI.6.14) with the unknowns δv^β, δk^β may be written [9]

$$D(\ell) \equiv (c^2 rf + \mu |h|^2) (u^\alpha \ell_\alpha)^2 - \mu (h^\alpha \ell_\alpha)^2$$

e) We have thus obtained three types of infinitesimal waves –or characteristic hypersurfaces of the main system of Magnetohydrodynamics.

1°) The ***entropy waves*** or hypersurfaces generated by the lines of flow which satisfy

$$u^\alpha \ell_\alpha = 0$$

At the traverse of an entropy wave front δS may be $\neq 0$

2°) ***The magnetosonic waves,*** hypersurfaces which satisfy

$$\text{(VI.6.15)}\qquad \begin{aligned} P(\ell) \equiv c^2 rf(u^\alpha \ell_\alpha)^4(\gamma-1) + (c^2 rf + \mu |h|^2 \gamma)(u^\alpha \ell_\alpha)^2 (\ell^\beta \ell_\beta) \\ - \mu(h^\alpha \ell_\alpha)^2 \ell^\beta \ell_\beta = 0 \end{aligned}$$

At the traverse of a magnetosonic wave front δp and consequently $\delta(u^\beta \ell_\beta)$, $\delta(h^\beta \ell_\beta)$ and $\delta |h|^2$ may be $\neq 0$.

3°) ***The Alfven waves*** satisfying

$$\text{(VI.6.16)}\qquad D(\ell) \equiv (c^2 rf + \mu |h|^2)(u^\alpha \ell_\alpha)^2 - \mu(h^\alpha \ell_\alpha)^2 = 0$$

At the traverse of an Alfven wave front, δv^β and δk^β may be $\neq 0$ for $\delta p = 0$.

Set $\beta = \sqrt{(c^2 rf + \mu |h|^2)/\mu}$. The equation $D(\ell) = 0$ may be written :

$$[(\beta u^\alpha + h^\alpha)\ell_\alpha][(\beta u^\alpha - h^\alpha)\ell_\alpha] = 0$$

and we see that the Alfven waves are generated by the integral curves of the vectors

$$A^\alpha = \beta u^\alpha + h^\alpha \qquad\qquad B^\alpha = \beta u^\alpha - h^\alpha$$

and we have two types of Alfven waves, the waves A and the waves B . We note that

$$A^\alpha A_\alpha = B^\alpha B_\alpha = \beta^2 - |h|^2 > 0$$

The vectors A and B are temporal and oriented toward the futur.

VI.7. - Magnetosonic speeds and Alfven speeds.

a) We have associated (see Section (V.1)) to a hypersurface Σ (of equation $\varphi = 0$, with $\ell = d\varphi$) , **u** being given, the parameter

$$y^\Sigma = (u^\alpha \ell_\alpha)^2 / \{(u^\alpha \ell_\alpha)^2 - \ell^\alpha \ell_\alpha\} \geqslant 0$$

This definition is equivalent to the relation

(VI.7.1) $$y^\Sigma / (1 - y^\Sigma) = -(u^\alpha \ell_\alpha)^2 / (\ell^\alpha \ell_\alpha)$$

Similarly associate with **h** a component h_n given by

(VI.7.2) $$h_n^2 = (h^\alpha \ell_\alpha)^2 / \{(u^\alpha \ell_\alpha)^2 - \ell^\alpha \ell_\alpha\} \geqslant 0$$

h_n^2 satisfies the following lemma :

Lemma 1°) *One has always* $h_n^2 \leqslant |h|^2$

2°) *One has* $h_n^2 = |h|^2$ *if and only if* ℓ *belongs to the 2-plane* π *defined by* u *and* h.

In fact, consider at the point n of Σ an orthonormalized basis $\{\ell_0, \ell_i\}$ $(i = 1, 2, 3)$ such that $\ell_0 = u$. One has $h^0 = 0$ and it follows from the Schwarz inequality

$$(h^i \ell_i)^2 \leqslant \Sigma (h^i)^2 \ \Sigma (\ell^i)^2$$

that is

$$(h^i \ell_i)^2 / \Sigma(\ell^i)^2 \leqslant \Sigma (h^i)^2 = |h|^2$$

let

$$h_n^2 \leqslant |h|^2$$

One has $h_n^2 = |h|^2$ if and only if, in the decomposition of ℓ along a vector colinear to u and a vector k orthogonal to u, k is colinear to h, that is to say that ℓ belongs to the 2-plane defined by u and h.

b) Substitute to $(h^\alpha \ell_\alpha)^2$ its expression in term of h_n^2 in $P(\ell)$. One has [9]

$$P(\ell) = c^2 rf(\gamma - 1)(u^\alpha \ell_\alpha)^4 + (c^2 rf + \mu |h|^2 \gamma - \mu h_n^2)(u^\alpha \ell_\alpha)^2 \ell^\beta \ell_\beta +$$

$$+ \mu h_n^2 (\ell^\beta \ell_\beta)^2$$

According to (VI.7.1), we may set

$$\text{(VI.7.3)} \qquad \frac{(y^{\Sigma})^2}{(u^{\alpha}\ell_{\alpha})^4}\, P(\ell) = \pi(y^{\Sigma})$$

where $\pi(y)$ denotes the trinome

$$\pi(y) = c^2rf(\gamma-1)\, y^2 - (c^2rf + \mu|h|^2\gamma - \mu h_n^2)\, y(1-y) + \mu h_n^2(1-y)^2$$

that is

$$\text{(VI.7.4)} \quad \pi(y) = (c^2rf + \mu|h|^2)\,\gamma y^2 - (c^2rf + \mu|h|^2\gamma + \mu h_n^2)\, y + \mu h_n^2$$

Under the hypothesis $\tau_p' < 0$ (that is $\gamma > 1$; see Section (V.2)), π has the following properties

$$\pi(0) = \mu h_n^2 \geqslant 0 \qquad\qquad \pi(1) = c^2rf(\gamma-1) > 0$$

But if v is the sonic speed of the fluid ($v^2 / c^2 = 1/\gamma$) one has

$$\pi(\frac{v^2}{c^2}) = \pi(\frac{1}{\gamma}) = (c^2rf + \mu|h|^2)\frac{1}{\gamma} - (c^2rf + \mu|h|^2\gamma + \mu h_n^2)\frac{1}{\gamma} + \mu h_n^2$$

let

$$\pi(\frac{v^2}{c^2}) = \mu(|h|^2 - h_n^2)(\frac{1}{\gamma} - 1) \leqslant 0$$

Therefore π admits two zeros between 0 and 1 , let y^{SM} and y^{RM}. We see that ***the equation of the magnetosonic waves defines two speeds*** v^{SM} ***and*** v^{RM} ***satisfying***

$$\text{(VI.7.5)} \qquad v^{SM} < v < v^{RM} < c$$

w^{SM} is the ***slow magnetosonic speed*** and v^{RM} ***the rapid magnetosonic speed***

c) If we introduce h_n^2 into $D(\ell)$, one has

(VI.7.6) $$D(\ell) = (c^2rf+\mu|h|^2)((u^\alpha \ell_\alpha)^2 - \ell^\alpha \ell_\alpha)(y^\Sigma - \frac{\mu h_n^2}{c^2rf + \mu|h|^2}) = 0$$

Therefore the Alfven waves admit with respect to the fluid a speed v^A given by

(VI.7.7) $$(v^A)^2 / c^2 = \frac{\mu h_n^2}{c^2rf + \mu|h|^2} < 1$$

Evaluate from the expression (VI.7.4) of π

$$\pi((v^A)^2 / c^2) = ((v^A)^2 / c^2)\, \mu\, (h_n^2 - |h|^2)\, (\gamma - 1) < 0$$

We see that we have

(VI.7.8) $$v^{SM} \leqslant v^A \leqslant v^{RM}$$

Suppose $u^\alpha \ell_\alpha \neq 0$. Let Σ be an Alfven wave with the speed v^A ($h_n^2 \neq 0$) : this wave satisfies $D(\ell) = 0$. We note that this wave satisfies also $P(\ell) = 0$ if and only if $\pi((v^A)^2/c^2) = 0$ that is if $h_n^2 = |h|^2$. According to the lemma, ℓ belongs to the 2-plane defined by u and h.

VI.8. - Representations of the wave cones in $\mathbb{R}^3$.

a) In each point x of the fluid, let T_x^* be the space of the covectors ℓ cotangent to the space-time. The main cone of the space-time may be defined by the cone Γ of covectors

$$\Gamma : g^{\alpha\beta} \ell_\alpha \ell_\beta = 0$$

It follows from our study that the characteristics (or infinitesimal waves) of magnetohydrodynamics are defined by the hypersurfaces tangent at one of the three cones

$$\Gamma_E : u^\alpha \ell_\alpha = 0 \qquad \Gamma_H : P(\ell) = 0 \qquad \Gamma_A : D(\ell) = 0$$

Consider at x an orthonormalized frame $(\mathbf{e}_0, \mathbf{e}_i)$ $(i = 1, 2, 3)$ such that

$$\mathbf{e}_0 = u \qquad \mathbf{e}_3 = h / |h|$$

We set in this Section :

$$\ell_0 = t \qquad \ell_1 = x \qquad \ell_2 = y \qquad \ell_3 = z$$

with these notations, we have for the different cones, the equations :

$$\Gamma : t^2 - x^2 - y^2 - z^2 = 0$$

$$\Gamma_E : t = 0$$

$$\Gamma_H : c^2 rf(\gamma - 1)t^4 + (c^2 rf + \mu |h|^2 \gamma)\, t^2 (t^2 - x^2 - y^2 - z^2)$$

$$- \mu |h|^2 z^2 (t^2 - x^2 - y^2 - z^2) = 0$$

$$\Gamma_A \, (c^2 rf + \mu |h|^2)\, t^2 - \mu |h|^2 z^2 = 0$$

with $\gamma > 1$. If we introduce $\beta^2 = (c^2 rf + \mu |h|^2) / \mu$ and $\overline{\beta^2} = \beta^2 / |h|^2$ the equations of Γ_H and Γ_A may be written

$$\begin{cases} \Gamma_H : (\overline{\beta^2} - 1)(\gamma - 1)t^4 + (\overline{\beta^2} + \gamma - 1)t^2 (t^2 - x^2 - y^2 - z^2) - \\ \qquad - z^2 (t^2 - x^2 - y^2 - z^2) = 0 \\ \Gamma_A : \overline{\beta^2}\, t^2 - z^2 = 0 \end{cases}$$

We call ***indicatrices in*** $\mathbb{R}^3$ the sections of these cones by the hyperplane $t = 1$. We obtain thus for Γ, Γ_M, Γ_A the three following indicatrices :

$$S : 1 - x^2 - y^2 - z^2 = 0$$

$$S_H : \overline{\beta^2}\, \gamma - (\overline{\beta^2} + \gamma - 1)(x^2 + y^2 + z^2) - z^2 (1 - x^2 - y^2 - z^2) = 0$$

$$S_A : z = \pm \overline{\beta}$$

Discuss the form of these indicatrices ; Oz is an axis of rotation and it is sufficient to cut the indicatrices by the plane $x = 0$. We obtain, for the sphere S, the circle

$$y^2 + z^2 = 1$$

for the two planes S_A , the two straight lines

$$z = \pm \overline{\beta}$$

and for S_M the quartic curve.

$$C_H : \ \overline{\beta}^2 \gamma - (\overline{\beta}^2 + \gamma - 1)(y^2 + z^2) - z^2(1 - y^2 - z^2) = 0$$

which has Oy and Oz as axis of symmetry. The equation of C_H may be written

$$y^2 = - \frac{(z^2 - \gamma)(z^2 - \overline{\beta}^2)}{z^2 - (\overline{\beta}^2 + \gamma - 1)}$$

For $z^2 = \gamma$ and $z^2 = \overline{\beta}^2$, one has $y^2 = 0$ and C_H admits the asymptoties $z^2 = \overline{\beta}^2 + \gamma - 1$.

b) We call ***general case*** the case where $\gamma \neq \overline{\beta}^2$. For example, for $\gamma < \overline{\beta}^2$, we have for C_H the form of Figure 1 ; Γ_H is topologically decomposed in two parts Γ_{H_1} and Γ_{H_2} . The exterior part Γ_{H_1} corresponds to the slow waves and the interior part Γ_{H_2} to the rapid waves.

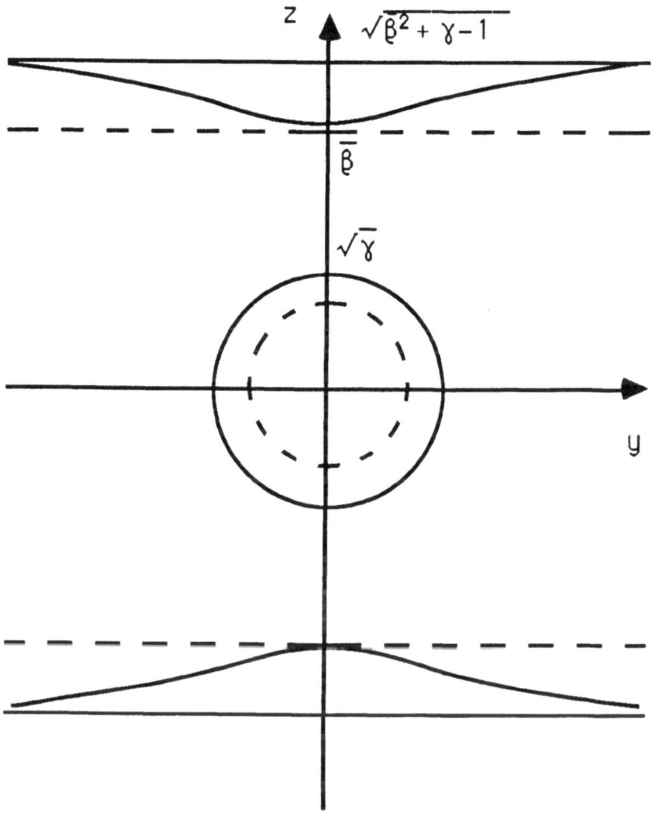

Figure 1

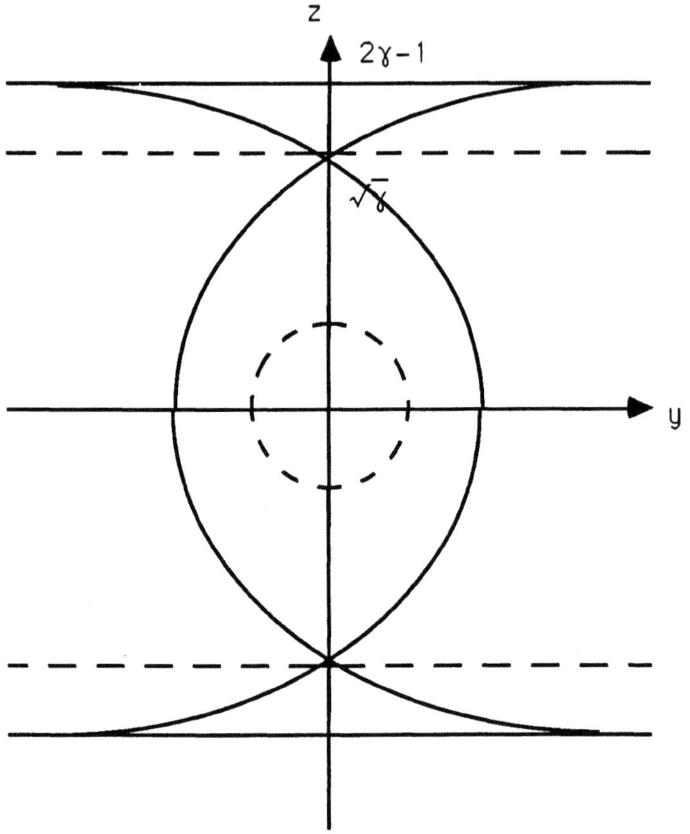

Figure 2

We call ***special case*** the case where $\gamma = \overline{\beta^2}$, that is [8]

$$\text{(VI.8.1)} \qquad \frac{v^2}{c^2} = \frac{\mu\,|h|^2}{c^2 r f + \mu\,|h|^2}$$

In this case, C_H admits two singular points on Oz and has the form of Figure 2. The cone Γ_H admits two singular generatrices $x = 0$, $y = 0$,

$z^2 = \gamma t^2$ belonging in the 2-plane Q defined by the vectors $\mathbf{u}$ and $\mathbf{h}$.

In all cases, the conic hypersurface Γ_A is the set of the two hyperplanes $z = \pm \overline{\beta} t$, the intersection of which is the 2-plane Q' defined by $z = 0$, $t = 0$, that is the 2-plane orthogonal to Q .

In ***the general case,*** the hyperplanes Γ_A are tangent to Γ_H along the generatrices $x = 0$, $y = 0$, $z = \pm \overline{\beta} t$. In the ***singular case,*** these hyperplanes are passing by these generatrices that are here singular.

The intersection of the hyperplane Γ_E $(t = 0)$ either with Γ_H or with each of the hyperplanes of Γ_A is nothing other but the 2-plane Q' .

c) Study now the intersection $\Gamma_M \cap \Gamma_A$. If we set $z^2 = \overline{\beta}^2 t^2$ in the equation of Γ_M , we have

$$(\overline{\beta}^2 - 1)(\gamma - 1)t^4 + (\overline{\beta}^2 + \gamma - 1)\, t^2 \; \{ t^2 (1 - \overline{\beta}^2 - x^2 - y^2 \}$$

$$- \overline{\beta}^2 t^2 \{ t^2 (1 - \overline{\beta}^2 - x^2 - y^2 \} = 0$$

that is to say

$$(\gamma - 1)\, t^2 (x^2 + y^2) = 0$$

The intersection is then the union of the 2-plane Q' $(z = 0 , t = 0)$ and of the two generatrices of Γ_H contained in the 2-plane Q and defined by $z = \pm \overline{\beta} t$.

d) Study the differential operators corresponding to $P(\ell)$ and $D(\ell)$. It follows from the previous considerations that ***in the general case,*** the operator corresponding to $P(\ell)$ is ***strictly hyperbolic ;*** it is not the same for the operator corresponding to $D(\ell)$ which is only the product of two strictly hyperbolic operators corresponding to the two hyperplanes.

The differential system of magnetohydrodynamics ***is not strictly hyperbolic.*** In the special case $\gamma = \overline{\beta}^2$, the operator corresponding to $P(\ell)$ itself is not strictly hyperbolic. In the general case, I have proved ([8]) a theorem of existence and uniqueness for the local Cauchy problem relative to the system of magnetohydrodynamics and to a spatial hypersurface over a suitable class of C^∞-functions (Gevrey class). This result has been generalized by Mrs Y. Choquet-Bruhat.

VI.9. - Main property of the rays associated to the magnetosonic waves.

a) Let φ be a solution of the equation corresponding to the magnetosonic wave fronts. We suppose that over $\Sigma\ (\varphi = 0)$ we have $D(\ell) \neq 0$, $u^\alpha \ell_\alpha \neq 0$. It follows from Sect. (VI.6) that the infinitesimal discontinuities δv^β, δk^β vanish and that $\delta(u^\beta \ell_\beta)$, $\delta(h^\beta \ell_\beta)$ are given in terms of δp. According to (VI.6.11), one has

$$\frac{1}{2}\,\mu \ell^\beta \ell_\beta\, \delta|h|^2 = c^2 r f(u^\alpha \ell_\alpha)\, \delta(u^\beta \ell_\beta) - (\ell^\alpha \ell_\alpha - (u^\alpha \ell_\alpha)^2)\, \delta p$$

that is, according to (VI.6.1)

$$\frac{1}{2}\mu \ell^{\beta}\ell_{\beta}\,\delta|h|^2 = -\{(\gamma-1)(u^{\alpha}\ell_{\alpha})^2 + \ell^{\alpha}\ell_{\alpha}\}\,\delta p$$

b) The magnetosonic fronts are the hypersurfaces tangent to the cones $P(\ell) = 0$. The generatrix of contact is defined by the vector

$$N^{\beta} = 1/2\,(\partial P(\ell)\,/\,\partial \ell_{\beta})$$

let :

$$N^{\beta} = 2c^2 rf(\gamma-1)(u^{\alpha}\ell_{\alpha})^3 u^{\beta} + (c^2 rf + \mu|h|^2\gamma)(u^{\alpha}\ell_{\alpha})(\ell^{\alpha}\ell_{\alpha}u^{\beta} +$$

$$+ (u^{\alpha}\ell_{\alpha})\ell^{\beta}) - \mu(h^{\alpha}\ell_{\alpha})(\ell^{\alpha}\ell_{\alpha}h^{\beta} + (h^{\alpha}\ell_{\alpha})\ell^{\beta}$$

The bicharacteristics or ***rays*** associated to the magnetosonic fronts Σ are the integral curves of the vector field N^{β} over Σ. We will show that δp ***is propagating along there rays*** according to a differential system

$$N^{\beta}\nabla_{\beta}\,\delta p \simeq 0$$

where the symbol $\simeq$ means terms proportional to δp.

We adopt the hypothesis of Section (I.6). The metric is supposed of class C^2 ; p, S, u^{α}. h^{α} are continuous over the domain Ω and of class C^2 over Ω^+ and Ω^- and their first and second derivatives are regularly discontinuous at the traverse of Σ. From (I.6.6) it follows that there exist distributions $\overline{p}$, $\overline{S}$, $\overline{u}^{\lambda}$ $\overline{h}^{\lambda}$ with supports over Σ such that

(VI.9.1) $$\overline{\delta}[\nabla_{\alpha}\nabla_{\beta}p] = \nabla_{\alpha}\ell_{\beta}\,\delta p + \ell_{\alpha}\nabla_{\beta}\,\delta p + \ell_{\beta}\nabla_{\alpha}\,\delta p + \ell_{\alpha}\ell_{\beta}\,\overline{p}$$

(VI.9.2) $$\overline{\delta}[\nabla_{\alpha}\nabla_{\beta}S] = \ell_{\alpha}\ell_{\beta}\,\overline{S} \qquad (\delta S = 0)$$

(VI.9.3) $\bar{\delta}[\nabla_\alpha \nabla_\beta u^\lambda] = \nabla_\alpha \ell_\beta \delta u^\lambda + \ell_\alpha \nabla_\beta \delta u^\lambda + \ell_\beta \nabla_\alpha \delta u^\lambda + \ell_\alpha \ell_\beta \bar{u}^\lambda$

(VI.9.4) $\bar{\delta}[\nabla_\alpha \nabla_\beta h^\lambda] = \nabla_\alpha \ell_\beta \delta h^\lambda + \ell_\alpha \nabla_\beta \delta h^\lambda + \ell_\beta \nabla_\alpha \delta h^\lambda + \ell_\alpha \ell_\beta \bar{h}^\lambda$

c) From the relations (VI.6.1) and (VI.6.2) one deduces

(VI.9.5) $c^2 rf\, \bar{\delta}[\nabla_\beta \nabla_\alpha u^\alpha] + \gamma\, u^\alpha \bar{\delta}[\nabla_\beta \nabla_\alpha p] \simeq 0$

(VI.9.6) $\bar{\delta}[\nabla_\beta \nabla_\alpha S] = 0$

From the relations (VI.4.6), (VI.4.7), consequences of the Maxwell equations, one obtains by differentiation

$$\bar{\delta}[\nabla_\beta \nabla_\alpha h^\alpha] - u^\alpha u^\lambda \bar{\delta}[\nabla_\beta \nabla_\alpha h_\lambda] \simeq 0$$

and :

$$\frac{1}{2} u^\alpha \bar{\delta}[\nabla_\alpha \nabla_\beta |h|^2] + |h|^2 \delta[\nabla_\beta \nabla_\alpha u^\alpha] - h^\alpha u^\lambda \bar{\delta}[\nabla_\beta \nabla_\alpha h^\lambda] \simeq 0$$

Multiplying the first relation by h^β, the second relation by u^β, it follows by difference

(VI.9.7) $\frac{1}{2} u^\alpha u^\sigma \bar{\delta}[\nabla_\alpha \nabla_\beta |h|^2] + |h|^2 u^\beta \bar{\delta}[\nabla_\beta \nabla_\alpha u^\alpha]$

$$- h^\beta \bar{\delta}[\nabla_\beta \nabla_\alpha h^\alpha] \simeq 0$$

From (VI.5.6) it follows in a similar way :

(VI.9.8) $c^2 rf\, \bar{\delta}[\nabla_\beta \nabla_\alpha h^\alpha] + h^\alpha \bar{\delta}[\nabla_\beta \nabla_\alpha p] \simeq 0$

From (VI.9.7) it follows according to (VI.9.8) and (VI9.5)

$$\text{(VI.9.9)}\quad \frac{1}{2} c^2 rf\, u^\alpha u^\beta \bar{\delta}[\nabla_\alpha \nabla_\beta |h|^2] \simeq (|h|^2 \gamma\, u^\alpha u^\beta - h^\alpha h^\beta)\, \bar{\delta}[\nabla_\beta \nabla_\alpha p]$$

d) Take the covariant derivative of (VI.5.5). One obtains

$$(c^2 rf + \mu |h|^2)\, u^\beta \bar{\delta}[\nabla_\beta \nabla_\alpha u^\alpha] - (g^{\alpha\beta} - u^\alpha u^\beta)\, \bar{\delta}[\nabla_\alpha \nabla_\beta p]$$

$$-\frac{1}{2} \mu g^{\alpha\beta} \bar{\delta}[\nabla_\alpha \nabla_\beta |h|^2] + \mu u^\alpha u^\beta \bar{\delta}[\nabla_\alpha \nabla_\beta |h|^2] +$$

$$\mu |h|^2 u^\beta \bar{\delta}[\nabla_\beta \nabla_\alpha u^\alpha] - 2\mu h^\beta \bar{\delta}[\nabla_\beta \nabla_\alpha h^\alpha] \simeq 0$$

let :

$$c^2 rf\; u^\beta \bar{\delta}[\nabla_\beta \nabla_\alpha u^\alpha] - (g^{\alpha\beta} - u^\alpha u^\beta)\, \bar{\delta}[\nabla_\alpha \nabla_\beta p]$$

$$-\frac{1}{2} \mu \bar{\delta}[\nabla_\beta \nabla^\beta |h|^2] + \mu \{ 2|h|^2 u^\beta \bar{\delta}[\nabla_\beta \nabla_\alpha u^\alpha] +$$

$$u^\alpha u^\beta\; \bar{\delta}[\nabla_\alpha \nabla_\beta |h|^2] - 2 h^\beta \bar{\delta}[\nabla_\beta \nabla_\alpha h^\alpha] \simeq 0$$

where the second line vanishs, according to (VI.9.7). It follows from (VI.9.5) that

$$\text{(VI.9.10)}\quad \{(\gamma - 1) u^\alpha u^\beta + g^{\alpha\beta}\}\, \bar{\delta}(\nabla_\alpha \nabla_\beta p] + \frac{1}{2} \mu \bar{\delta}[\nabla_\beta \nabla^\beta |h|^2] \simeq 0$$

But one has

$$\frac{1}{2}\; \bar{\delta}[\nabla_\beta \nabla^\beta |h|^2] \simeq \ell_\beta \nabla_\beta \delta |h|^2 + \frac{1}{2} \ell^\rho \ell_\rho |\bar{h}|^2$$

and (VI.9.9) may be written

$$c^2 rf\,(u^\alpha \ell_\alpha)\, u^\beta \nabla_\beta\, \delta |h|^2 + \frac{1}{2}\, c^2 rf\,(u^\alpha \ell_\alpha)^2\, \overline{h}^2 \simeq$$

$$\simeq (\,|h|^2 \gamma\, u^\alpha u^\beta - h^\alpha h^\beta\,)\, \overline{\delta}\,[\nabla_\alpha \nabla_\beta\, p\,]$$

Therefore after multiplication by $c^2 rf\,(u^\rho \ell_\rho\,)^2$, the formula (VI.9.10) takes the form

$$\{c^2 rf\,((\gamma-1)\, u^\alpha u^\beta + g^{\alpha\beta}\}\,(u^\rho \ell_\rho)^2\, \overline{\delta}\,[\nabla_\alpha \nabla_\beta\, p\,] +$$

$$\text{(VI.9.11)} \qquad + \mu\, c^2 rf\,(u^\rho \ell_\rho)^2\, \ell^\beta \nabla_\beta\, \delta |h|^2 + \mu\, \ell^\rho \ell_\rho \times$$

$$\times \{-c^2 rf\,(u^\alpha \ell_\alpha)\, u^\beta \nabla_\beta\, \delta |h|^2 + (\,|h|^2 \gamma\, u^\alpha u^\beta - h^\alpha h^\beta\,)\}\, \overline{\delta}\,[\nabla_\alpha \nabla_\beta\, p\,] \simeq 0$$

let

$$\{c^2 rf\,((\gamma-1)\, u^\alpha u^\beta + g^{\alpha\beta}\,)\,(u^\rho \ell_\rho)^2 + \mu |h|^2 \gamma\, \ell^\rho \ell_\rho\, u^\alpha u^\beta - \mu\, \ell^\rho \ell_\rho\, h^\alpha h^\beta\}\times$$

(VI.9.12)

$$\times\, \overline{\delta}\,[\nabla_\alpha \nabla_\beta\, p\,] + \mu\, c^2 rf\,(u^\rho \ell_\rho)\, \{(u^\rho \ell_\rho)\, \ell^\beta - \ell^\rho \ell_\rho\, u^\beta\}\, \nabla_\beta\, \delta |h|^2 \simeq 0$$

Substituting to $\overline{\delta}\,[\nabla_\alpha \nabla_\beta\, p\,]$ its value (VI.9.1), the coefficient of $\overline{p}$ is

$$c^2 rf(\gamma-1)(u^\alpha \ell_\alpha)^4 + (c^2 rf + \mu |h|^2 \gamma)\, \ell^\rho \ell_\rho (u^\alpha \ell_\alpha)^2 -$$

$$- \mu\, \ell^\rho \ell_\rho (h^\alpha \ell_\alpha)^2 = P(\ell) = 0$$

We deduce from (VI.9.12)

$$\{ c^2 rf(\gamma-1)(u^\alpha \ell_\alpha)^3 u^\beta + c^2 rf (u^\alpha \ell_\alpha)^2 \ell^\beta +$$

(VI.9.13) $$+ \mu |h|^2 \gamma \ell^\rho \ell_\rho (u^\alpha \ell_\alpha) u^\beta - \mu(\ell^\rho \ell_\rho)(h^\alpha \ell_\alpha) h^\beta \} \nabla_\beta \delta p +$$

$$+ \frac{1}{2} \mu c^2 rf (u^\rho \ell_\rho) \{(u^\rho \ell_\rho) \ell^\beta - (\ell^\rho \ell_\rho) u^\beta\} \nabla_\beta \delta |h|^2 \simeq 0$$

The vector coefficient of $\nabla_\beta \delta p$ and the vector coefficient of $\nabla_\beta \delta |h|^2$ are tangent to Σ and one has

$$\{c^2 rf(\gamma-1)(u^\alpha \ell_\alpha)^3 v^\beta + \mu |h|^2 \gamma \ell^\rho \ell_\rho (u^\alpha \ell_\alpha) v^\beta - \mu(\ell^\rho \ell_\rho)(h^\alpha \ell_\alpha) k^\beta \} \nabla_\beta \delta p$$

(VI.9.14)

$$- \frac{1}{2} \mu c^2 rf \, \ell^\rho \ell_\rho (u^\alpha \ell_\alpha) v^\beta \, \nabla_\beta \delta |h|^2 \simeq 0$$

let according to (VI.9.1)

$$\{2c^2 rf(\gamma-1)(u^\alpha \ell_\alpha)^3 v^\beta + (c^2 rf + \mu |h|^2 \gamma) \ell^\rho \ell_\rho (u^\alpha \ell_\alpha) v^\beta -$$

$$- \mu \ell^\rho \ell_\rho (h^\alpha \ell_\alpha) k^\beta \} \nabla_\beta \delta p \simeq 0$$

But N^β being tangent to Σ , one has

$$N^\beta = 2c^2 rf(\gamma-1)(u^\alpha \ell_\alpha)^3 v^\beta + (c^2 rf + \mu |h|^2 \gamma) (u^\alpha \ell_\alpha) \ell^\rho \ell_\rho v^\beta -$$

$$- \mu \ell^\rho \ell_\rho (h^\alpha \ell_\alpha) k^\beta$$

Therefore

$$N^\beta \nabla_\beta \delta p \simeq 0$$

Theorem. - *Under the hypotheses of* Section (VI.9) b *the infinitesimal discontinuities* δp , $\delta(u^\alpha \ell_\alpha)$, $\delta(h^\alpha \ell_\alpha)$ *associated to a magnetosonic wave are propagating along the corresponding rays according to*

$$N^\beta \nabla_\beta \delta p \simeq 0 \qquad N^\beta \nabla_\beta \delta(u^\alpha \ell_\alpha) \simeq 0 \qquad N^\beta \nabla_\beta \delta(h^\alpha \ell_\alpha) \simeq 0$$

VI.10. - Property of the rays associated to the Alfven waves.

a) Let φ be a solution of the equation of the Alfven waves, of type A for example.

One has

$$A^\alpha \partial_\alpha \varphi = A^\alpha \ell_\alpha = (\beta u^\alpha + h^\alpha)\ell_\alpha = 0 \tag{VI.10.1}$$

We suppose that $P(\ell) \neq 0$, $u^\alpha \ell_\alpha \neq 0$. It follows

$$\delta p = 0 \qquad \delta(u^\alpha \ell_\alpha) = 0 \qquad \delta(h^\alpha \ell_\alpha) = 0 \qquad \delta |h|^2 = 0 \qquad \delta S = 0$$

From Section (VI.9) one deduces easily

$$\bar{\delta}[\nabla_\alpha \nabla_\alpha p] = 0 \ , \ \bar{\delta}[\nabla_\beta \nabla_\alpha u^\alpha] = 0 \ , \ \bar{\delta}[\nabla_\beta \nabla_\alpha h^\alpha] = 0$$

(VI.10.2)

$$\bar{\delta}[\nabla_\alpha \nabla_\beta |h|^2] = 0 \qquad , \qquad \bar{\delta}[\nabla_\alpha \nabla_\beta S] = 0$$

b) Here only δv^λ , δk^λ may be non vanishing. It follows from (VI.6.13) , (VI.6.14) that δk^λ is proportional to δv^λ . The symbol $\simeq$ means in this Section modulo linear terms in δv^λ (or in δk^λ) .

From (VI.10.2), we deduce from the Maxwell equations (VI.4.5) and from the system (VI.5.5) of the lines of flow

$$h^\alpha \bar{\delta} [\nabla_\alpha \nabla_\beta u^\lambda] - u^\alpha \bar{\delta} [\nabla_\alpha \nabla_\beta h^\lambda] \simeq 0$$

and

$$\beta^2 u^\alpha \bar{\delta} [\nabla_\alpha \nabla_\beta u^\lambda] - h^\alpha \bar{\delta} [\nabla_\alpha \nabla_\beta h^\lambda] \simeq 0$$

Multplying the first relation by h^β , the second relation by u^β , one obtains by difference

$$(\beta^2 u^\alpha u^\beta - h^\alpha h^\beta) \bar{\delta} [\nabla_\alpha \nabla_\beta u^\lambda] \simeq 0$$

let according to (VI.9.3)

$$(\beta^2 u^\alpha u^\beta - h^\alpha h^\beta) (\ell_\alpha \nabla_\beta \delta u^\lambda + \ell_\beta \nabla_\alpha \delta u^\lambda + \ell_\alpha \ell_\beta \bar{u}^\lambda) \simeq 0$$

The coefficient of $\bar{u}^\lambda$ is null and one has

$$\{ \beta^2 (u^\alpha \ell_\alpha) u^\beta - (h^\alpha \ell_\alpha) h^\beta \} \nabla_\beta \delta u^\lambda \simeq 0$$

that is, according to (IV.10.1)

$$\beta (u^\alpha \ell_\alpha) A^\beta \nabla_\beta \delta u^\lambda \simeq 0$$

Therefore [9]

$$A^{\beta} \nabla_{\beta} \delta u^{\lambda} \simeq 0$$

Theorem : *The distributions* δv^{λ} , δk^{λ} *with supports over an Alfven wave front* Σ *of type* A *are propagating along the associated rays according to*

$$A^{\beta} \nabla_{\beta} \delta v^{\lambda} \simeq 0 \qquad\qquad A^{\beta} \nabla_{\beta} \delta k^{\lambda} \simeq 0$$

where $\simeq$ *means modulo linear terms in* δv^{μ} *(or* δk^{μ}*)*

Similar results are available for an Alfven wave of type B .

c) For a wave of type A , study the action of the derivation δ on the vector

$$A^{\beta} = \beta u^{\beta} + h^{\beta}$$

It follows from the study of a)

$$\delta\beta = 0 \qquad\qquad \delta(u^{\alpha} \ell_{\alpha}) = 0 \qquad\qquad \delta(h^{\alpha} \ell_{\alpha}) = 0$$

and one has

$$\delta A^{\beta} = \beta \delta v^{\beta} + \delta k^{\beta}$$

According to (VI.6.13) where the right side is null, we have

$$(u^{\alpha} \ell_{\alpha}) \ \delta A^{\beta} = (\beta \, (u^{\alpha} \ell_{\alpha}) + h^{\alpha} \ell_{\alpha}) \ \delta v^{\beta} = 0$$

One obtains $\delta A^{\beta} = 0$. *The vector* A^{β} *itself is invariant under the derivation* δ.

VII. - MAGNETOHYDRODYNAMIC SHOCK WAVES.

VII.1. - The main system of the shock waves.

a) A state Y of the system (fluid + field) at the point x of V_4 is defined by the values of p , S , $\mathbf{u}$, $\mathbf{h}$ (eight parameters). Over a domain Ω of V_4 , a magnetohydrodynamic shock wave is a solution of the fundamental system in a weak sense such that there exists a hypersurface Σ , the wave front, satisfying the following conditions

1) *On both sides of* Σ *, that is over* Ω^+ *and* Ω^- *, the states are* C^1 *- functions of* x *and the fundamental system is satisfied in the usual sense.*

2) *The variables defining the states are regularly discontinuous at the traverse of* Σ *, and in the neighborhood of* Σ *, the fundamental system is satisfied in the sense of distributions.*

We denote by Y_0 the state at $x \in \Sigma$ before the shock and by Y_1 the state at x after the shock ; $[Q] = Q_1 - Q_0$ is the discontinuity of a quantity Q at the traverse of Σ .

Over Ω^+ (resp. Ω^-) , one has

$$\nabla_\alpha (r u^\alpha) = 0 \qquad \nabla_\alpha (h^\alpha u^\beta - u^\alpha h^\beta) = 0 \qquad \nabla_\alpha T^{\alpha\beta} = 0$$

Over Ω , we adopt as fundamental system in a weak sense, the following system

$$\nabla_\alpha (r\,u^\alpha)^D = 0 \qquad \nabla_\alpha (h^\alpha u^\beta - u^\alpha h^\beta)^D = 0 \qquad \nabla_\alpha (T^{\alpha\beta})^D = 0$$

and our hypotheses imply :

$$(\nabla_\alpha (r\,u^\alpha))^D = 0 \qquad (\nabla_\alpha (h^\alpha u^\beta - u^\alpha h^\beta))^D = 0 \qquad (\nabla_\alpha (T^{\alpha\beta}))^D = 0$$

It follows from (I.4.5)

$$\text{(VII.1.1)} \quad \bar{\delta}[\,r\,u^\alpha \ell_\alpha\,] = 0 \quad \bar{\delta}[(h^\alpha u^\beta - u^\alpha h^\beta)\ell_\alpha\,] = 0 \quad \bar{\delta}[\,T^{\alpha\beta}\ell_\alpha\,] = 0$$

Set : ([1] and [9])

$$a(Y) = r\,u^\alpha \ell_\alpha \qquad V^\beta(Y) = (h^\alpha \ell_\alpha)\,u^\beta - \frac{a(Y)}{r}\,h^\beta$$

and

$$W^\beta(Y) = (c^2\tau + \mu\,\frac{|h|^2}{r^2})\;\;a(Y)\,r\,u^\beta - q\,\ell^\beta - \mu\,(h^\alpha \ell_\alpha)\,h^\beta$$

where $\mathbf{V}^\beta$ is tangent to Σ. The system (VII.1.1) expresses that :

$$a(Y_1) = a(Y_0) \qquad V^\beta(Y_1) = V^\beta(Y_0) \qquad W^\beta(Y_1) = W^\beta(Y_0)$$

The scalar a and the vectors $\mathbf{V}^\beta$ and W^β define ***the shock invariants.***

b) We say that a shock is ***tangential*** if $a = 0$. If such is the case :

$$u_0^\alpha \ell_\alpha = u_1^\alpha \ell_\alpha = 0$$

Σ has a null speed with respect to the fluid in the two states and is oriented in the time. From the invariance of V^β and W^β, it follows

$$(\text{VII.1.2}) \qquad (h_1^\alpha \ell_\alpha)\, u_1^\beta = (h_0^\alpha \ell_\alpha)\, u_0^\beta$$

$$(\text{VII.1.3}) \qquad (q_1 - q_0)\, \ell^\beta + \mu\, \{ (h_1^\alpha \ell_\alpha)\, h_1^\beta - (h_0^\alpha \ell_\alpha)\, h_0^\beta \} = 0$$

If $h_0^\alpha \ell_\alpha \neq 0$, u_1^β and u_0^β being unitary, colinear according to (VII.1.2) and oriented towards the future, one has $u_1^\beta = u_0^\beta$ and consequently $h_1^\alpha \ell_\alpha = h_0^\alpha \ell_\alpha$. From (VII.1.3), it follows :

$$\ell^\beta \ell_\beta (q_1 - q_0) + \mu\, \{(h_1^\alpha \ell_\alpha)^2 - (h_0^\alpha \ell_\alpha)^2\} = 0$$

We obtain $[q] = 0$. We deduce from (VII.1.3) $[h^\beta] = 0$ and $[p] = 0$. The discontinuity of S is indetermined.

If $h_0^\alpha \ell_\alpha = 0$, it follows from (VII.1.2) $h_1^\alpha \ell_\alpha = 0$. The shock equations give only :

$$u_1^\alpha \ell_\alpha = u_0^\alpha \ell_\alpha = 0 \qquad h_1^\alpha \ell_\alpha = h_0^\alpha \ell_\alpha = 0 \qquad q_1 = q_0$$

the other discontinuities being indetermined.

We do not consider in the following this trivial case and suppose $a \neq 0$ ***(non tangential shock).***

VII.2. - Analysis of the shock equations

a) Introduce the scalar H , invariant under the shock, given by

(VII.2.1) $$H(Y) = (1/a^2)\ V^\beta V_\beta = \frac{(h^\alpha \ell_\alpha)^2}{a^2} - \frac{|h|^2}{r^2}$$

Express h^β in terms of u^β and V^β :

$$h^\beta = \frac{h^\alpha \ell_\alpha}{a}\, r\, u^\beta - \frac{r}{a}\, V^\beta$$

We obtain for W^β (suppressing the reference to Y)

$$W^\beta = (c^2 \tau + \mu \frac{|h|^2}{r^2})\, a\, r\, u^\beta - q \ell_\beta - \mu \frac{(h^\alpha \ell_\alpha)^2}{a}\, r\, u^\beta + \mu \frac{r(h^\alpha \ell_\alpha)}{a}\, V^\beta$$

let

(VII.2.2) $$W^\beta = a\, r\, \alpha\, u^\beta - q \ell^\beta + \mu \frac{r}{a} (h^\alpha \ell_\alpha)\, V^\beta$$

where we have introduced the important variable :

(VII.2.3) $$\alpha = c^2 \tau - \mu H = D(\ell) / a^2$$

Decompose W^β between its tangential part and its transverse part with respect to Σ . One has *for* $\ell^\alpha \ell_\alpha \neq 0$

$$W^\beta = X^\beta - (q - \frac{a^2 \alpha}{\ell^\alpha \ell_\alpha})\, \ell^\beta$$

where

(VII.2.4) $$X^\beta = a\, r\, \alpha\, v^\beta + \mu \frac{r}{a} (h^\alpha \ell_\alpha)\, V^\beta$$

is tangent to Σ . The invariance of W^{β} is equivalent to the invariance of X^{β} and of that the traverse component. Therefore the scalar :

$$e = q - (c^2 a^2 / (\ell^{\alpha} \ell_{\alpha}))\ \tau \qquad (\ell^{\alpha} \ell_{\alpha} \neq 0)$$

is invariant under the shock.

Consider in particular the invariant scalar product

$$W^{\beta} v_{\beta} = a\, r\, (h^{\alpha} \ell_{\alpha})\, (\alpha + \mu\ H)$$

According to the definition of α , one obtains $W^{\beta B} v_{\beta} = c^2\, a\, b$, where b is the invariant scalar :

(VII.2.5) $$b = f\, (h^{\alpha} \ell_{\alpha})$$

b) Consider the invariant scalar $(\ell^{\alpha} \ell_{\alpha} \neq 0)$

$$K = (1/c^2 a^2)\ X^{\beta}\ X_{\beta}$$

We will prove

Lemma 1. - *The invariant* K *may be written*

(VII.2.6) $$K = c^2 f^2 - (c^2 a^2 / \ell^{\alpha} \ell_{\alpha})\ \tau^2 + 2\, \mu\, \tau\, \chi - (\mu^2 H / c^2)\ \chi$$

where

$$\chi = |\, h\, |^2 + (a^2 / \ell^{\alpha} \ell_{\alpha})\, H$$

K *admits also the expression*

(VII.2.7) $$K = c^2 f^2 + \mu |h|^2 \tau + (\mu |h|^2 / c^2)\, \alpha - (a^2/c^2 (\ell^\alpha \ell_\alpha))\, \alpha^2$$

From

$$r\, v^\beta = r\, u^\beta - (a / (\ell^\alpha \ell_\alpha))\, \ell^\beta$$

we deduce

$$(r\, v^\beta)(r\, v_\beta) = r^2 - a^2/(\ell^\alpha \ell_\alpha) \qquad (r\, v^\beta)\, \mathbf{v}_\beta = (r\, u^\beta)\, \mathbf{v}_\beta = r\, (h^\alpha \ell_\alpha)$$

It follows

(VII.2.8) $$K = (r^2 - a^2/(\ell^\alpha \ell_\alpha))\, (\alpha^2/c^2) + 2(\mu\alpha/c^2 a^2)\, r^2 (h^\alpha \ell_\alpha)^2 + \\ + (\mu^2/c^2 a^2)\, r^2 (h^\alpha \ell_\alpha)^2 H$$

Substituing the value of α , one has

$$K = c^2 (r^2 - a^2/(\ell^\alpha \ell_\alpha))(\tau - \mu H/c^2)^2 + 2\mu (r^2 (h^\alpha \ell_\alpha)^2/a^2)(\tau - \mu H/c^2) + \\ + (\mu^2/c^2 a^2)\, r^2 (h^\alpha \ell_\alpha)^2 H$$

let

$$K = c^2 (r^2 - a^2/(\ell^\alpha \ell_\alpha))\, \tau^2 + 2\mu\tau\, \{(r^2 (h^\alpha \ell_\alpha)^2/a^2) - \\ (r^2 - a^2/(\ell^\alpha \ell_\alpha))\, H\} - (\mu^2 H/c^2)\{(r^2 (h^\alpha \ell_\alpha)^2 (a^2) - (r^2 - a^2/(\ell^\alpha \ell_\alpha))\, H\}$$

But according to the definition of H :

$$\chi = r^2(h^2 \ell_\alpha)^2 / a^2 - (r^2 - a^2 / (\ell^\alpha \ell_\alpha)) H = r^2 (h^\alpha \ell_\alpha)^2 / a^2 -$$

$$- r^2 ((h^\alpha \ell_\alpha)^2 / a^2 - |h|^2 / r^2) + a^2 H / (\ell^\alpha \ell_\alpha)$$

let

$$\chi = |h|^2 + a^2 H / (\ell^\alpha \ell_\alpha)$$

It follows

$$K = c^2 f^2 - c^2 a^2 \tau^2 / (\ell^\alpha \ell_\alpha) + 2 \mu \tau \chi - \mu^2 H \chi / c^2$$

that is (VII.2.6). We deduce from this relation :

$$K = c^2 f^2 - c^2 a^2 \tau^2 / (\ell^\alpha \ell_\alpha) + 2 \mu \tau \ (|h|^2 + a^2 H / (\ell^\alpha \ell_\alpha))$$

$$- (\mu^2 H / c^2)(|h|^2 + a^2 H / (\ell^\alpha \ell_\alpha))$$

let

$$K = c^2 f^2 + \mu |h|^2 (2\tau - \mu H / c^2) -$$

$$- (a^2 / c^2 (\ell^\alpha \ell_\alpha))(c^4 \tau^2 - 2 c^2 \tau \mu H + \mu^2 H^2)$$

In terms of α , one obtains (VII.2.7)

c) Transform the expression of HK , where K is given by (VII.2.8).

One has

$$HK = H (r^2 - a^2 / (\ell^\alpha \ell_\alpha)) (\alpha^2 / c^2) +$$

$$+ (r^2 (h^\alpha \ell_\alpha)^2 / c^2 a^2) \mu H (2\alpha + \mu H)$$

But $\mu H = c^2 \tau - \alpha$ and one obtains

$$HK = H\,(r^2 - a^2 /(\ell^\alpha \ell_\alpha))\,(\alpha^2 / c^2) +$$

$$+ (r^2 (h^\alpha \ell_\alpha)^2 / c^2 a^2)\,(c^2 \tau + \alpha)\,(c^2 \tau - \alpha)$$

let

$$HK = H\,(r^2 - a^2 /(\ell^\alpha \ell_\alpha))\,(\alpha^2 / c^2) +$$

$$+ (r^2 (h^\alpha \ell_\alpha)^2 / c^2 a^2)\,(c^4 \tau^2 - \alpha^2)$$

It follows

$$HK = \{ H\,(r^2 - a^2 (\ell^\alpha \ell_\alpha)) - r^2 (h^\alpha \ell_\alpha)^2 / a^2)\}\,(\alpha^2 / c^2)$$

$$+ c^2 f^2 (h^\alpha \ell_\alpha)^2 / a^2$$

let

$$HK = c^2 b^2 / a^2 - \chi\, \alpha^2 / c^2 \qquad \text{(VII.2.9)}$$

We see that $L = \chi\, \alpha^2$ is an invariant under the shock. The invariants a, b, H, KL are not independant.

d) Concerning χ , we will prove the following lemma.

Lemma 2. 1) *One has* $(\ell^\alpha \ell_\alpha)\ \chi \leqslant 0$

2) *If* $\ell^\alpha \ell_\alpha$ *is* $\neq 0$, $\chi = 0$ *if and only if* ℓ *belongs to the 2-plane* π *defined by* (u , h)

3) *If* $\ell^\alpha \ell_\alpha$ *is* $\geqslant 0$ *, one has* $H \leqslant 0$ *and thus* $\alpha > 0$

According to the definitions of χ and H, one has

$$(\ell^\alpha \ell_\alpha)\,\chi = (\ell^\alpha \ell_\alpha)\,|h|^2 + (h^\alpha \ell_\alpha)^2 - |h|^2 (u^\alpha \ell_\alpha)^2$$

let, in terms of h_n^2,

$$(\ell^\alpha \ell_\alpha)\,\chi = \{(u^\alpha \ell_\alpha)^2 - (\ell^\alpha \ell_\alpha)\}\,(h_n^2 - |h|^2)$$

and according to the lemma of Section (VI.7), 1 and 2 are proved.

The inequality $h_n^2 \leqslant |h|^2$ may be written

$$(h^\alpha \ell_\alpha)^2 \leqslant |h|^2 \,((u^\alpha \ell_\alpha)^2 - \ell^\alpha \ell_\alpha)$$

let

$$\frac{(h^\alpha \ell_\alpha)^2}{a^2} \leqslant \frac{|h|^2}{r^2} - \frac{|h|^2}{a^2}\,\ell^\alpha \ell_\alpha$$

it follows

$$H = (h^\alpha \ell_\alpha)^2 / a^2 - |h|^2 / r^2 \leqslant -|h|^2 (\ell^\alpha \ell_\alpha) / a^2$$

For $\ell^\alpha \ell_\alpha \geqslant 0$, one has $H \leqslant 0$ and thus $\alpha = c^2\tau - \mu H > 0$

VII. 3. - New form of the shock equations for $\ell^\alpha \ell_\alpha \neq 0$

a) From the shock equations (VII.1.1), it follows that the two thermodynamical variables and the three scalars $|h|^2$, $u^\alpha \ell_\alpha$, $h^\alpha \ell_\alpha$, satisfy five fundamental relations of invariance. Moreover ***we add to*** (VII.1.1) ***the hypothesis***

(VII.3.1) $$[S] \geqslant 0$$

which is the Clausius-Duhem inequality

We may take as fundamental relations of invariance [9]

(VII.3.2) $$r\, u^\alpha \ell_\alpha = a$$

(VII.3.3) $$f\, h^\alpha \ell_\alpha = b$$

(VII.3.4) $$(h^\alpha \ell_\alpha)^2 / a^2 - |h|^2 / r^2 = H$$

(VII.3.5) $$\chi\, \alpha^2 = L \qquad (\ell^\alpha \ell_\alpha \neq 0)$$

(VII.3.6) $$\bar{q} - (c^2 a^2 / (\ell^\alpha \ell_\alpha))\, \tau = \bar{e} \qquad (\ell^\alpha \ell_\alpha \neq 0)$$

where we have set :

$$\alpha = c^2 \tau - \mu H \qquad \chi = |h|^2 + (a^2 / (\ell^\alpha \ell_\alpha)) H$$

$$\bar{q} = p + \frac{1}{2} \mu \chi = q + \frac{1}{2} \mu (a^2 / (\ell^\alpha \ell_\alpha))\, H$$

We note that, for $H \neq 0$, these relations imply, according to (VII.2.9), that K is invariant under the shock.

We shall prove that, under suitable conditions, this system of invariance relations defines in a unique way the values after the shock of the two thermodynamic variables and of the three scalars $(h_1)^2$, $u_1^\alpha \ell_\alpha$, $h_1^\alpha \ell_\alpha$. It is clear that if (VII.3.4), (VII.3.5), (VII.3.6) define the values of χ (or $|h|^2$) and of the thermodynamic variables, (VII.3.2), (VII.3.3) give in a unique way the transverse values of $\mathbf{u}$ and $\mathbf{h}$.

b) Consider now the tangential components v^β of the velocity and k^β of the magnetic field ; $\mathbf{V}^\beta$ and χ^β being invariant, one has

(VII.3.7) $$(h_1^\alpha \ell_\alpha) v_1^\beta - (u_1^\alpha \ell_\alpha) k_1^\beta = (h_0^\alpha \ell_\alpha) v_0^\beta - (u_0^\alpha \ell_\alpha) k_0^\beta$$

(VII.3.8) $$(c^2 r_1 f_1 + \mu |h_1|^2)(u_1^\alpha \ell_\alpha) v_1^\beta - (h_1^\alpha \ell_\alpha) k_1^\beta = (c^2 r_0 f_0 + \mu |h_0|^2)$$
$$(u_0^\alpha \ell_\alpha) v_0^\alpha - (h_0^\alpha \ell_\alpha) k_0^\beta$$

The determinant of the left-hand sides of (VII.3.7), (VII.3.8) with respect to v_1^β, k_1^β is $D_1(\ell) = a^2 \alpha_1$. If $\alpha_1 \neq 0$, (VII.3.7), (VII.3.8) give v_1^β, k_1^β in terms of the initial state and quantities that satisfy the five scalar invariance relations (VII.3.2)-(VII.3.6).

Suppose $\alpha_1 = 0$; Σ is then a front of an Alfven wave, oriented in time after the shock ($\ell^\alpha \ell_\alpha < 0$). It follows from (VII.3.5) that either $\alpha_0 = 0$ or $\chi_0 = 0$. One may establish the following result (Ann. 1) : a shock wave such that $\alpha_1 \alpha_0 = 0$ is compatible with ordinary infinitesimal waves only if it is ***an Alfven shock*** $(\alpha_1 = \alpha_0 = 0)$; The cases $\alpha_1 = 0$, $\chi_0 = 0$, $\alpha_0 \neq 0$ and symmetrically $\alpha_0 = 0$, $\chi_1 = 0$, $\alpha_1 \neq 0$, (singular shocks) are ***physically forbidden as unstable*** with respect to the infinitesimal Alfven waves.

VII.4. - Alfven shocks.

a) We have seen that we have ***an Alfven shock*** if $\alpha_9 = \alpha_0 = 0$. We will prove the following.

Theorem 1) ***If in a shock, we have*** $\alpha_1 = \alpha_0 \neq 0$, *the shock is null*

2) *In an Alfven shock* $(\alpha_1 = \alpha_0 = 0)$, *the thermodynamic variables and the three scalars* $|h|^2$, $(u^\alpha \ell_\alpha)$, $(h^\alpha \ell_\alpha)$ *are continuous if the hypothesis* $\tau'_p < 0$ *is satisfied.*

1) If $[\alpha] = 0$, one has

$$[\tau] = 0$$

Under our hypothesis, one obtains

$$[\chi] = 0 \qquad\qquad [|h|^2] = 0$$

According to (VII.3.6), one has $[\bar{q}] = 0$ and thus

$$[p] = 0$$

In K given by (VII.2.7), the three last terms are invariant under the shock. One has then

$$[f] = 0 \qquad\qquad [r] = 0$$

and according to (VII.3.2) and (VII.3.3), one has

$$[u^\alpha \ell_\alpha] = 0 \qquad\qquad [h^\alpha \ell_\alpha] = 0$$

Σ being not Alfven wave, it follows from (VII.3.7) and (VII.3.8) that the shock is null.

2) For an Alfven shock, one has always $[\tau] = 0$ and it follows from the shock equations, K being given by (VII.2.7)

(VII.4.1) $$[p + \frac{1}{2}\mu |h|^2] = 0 \qquad\qquad [c^2 f^2 + \mu |h|^2 \tau] = 0$$

We deduce from (VII.4.1):

$$[c^2 r f + \mu |h|^2] = 0$$

According to $[\mu |h|^2] = -[2p]$, one has

(VII.4.2) $$[c^2 r f - 2p] = 0$$

Study the independance of the two thermodynamics variables τ and

$$\varphi = c^2 r f - 2p$$

One has

$$\varphi'_p = c^2 f r'_p - 1 = \gamma - 1 = - c^2 r^2 \tau'_p$$

In a similar way

$$\varphi'_S = c^2 f r'_S + r \Theta = r \ (c^2 f r'_s / r + \Theta)$$

But

$$c^2 \tau'_S = \Theta V + c^2 f \ v'_S$$

and thus

$$c^2 r \tau'_S = \Theta - c^2 f r'_S / r$$

It follows

$$\varphi'_S = r \, (2 \Theta - c^2 r \tau'_S)$$

The Jacobian of φ and τ with respect to p , S has as value

$$\frac{d(\varphi , \tau)}{d(p, S)} = - c^2 r^2 \tau'_p \tau'_S - r \, (2 \Theta - c^2 r \tau'_S) \ \tau'_p$$

let

$$\frac{d(\varphi , \Sigma)}{d(p, S)} = - 2 r \Theta \tau'_p$$

which is $\neq 0$ under the hypothesis $\tau'_p < 0$. The variables φ and τ are independant and continuous at the traverse of Σ . All the thermodynamic variables are continuous ; in particular $[r] = 0$, $[f] = 0$, $[p] = 0$; according to (VII.3.2) , (VII.3.3) , one has ;

$$[u^\alpha \ell_\alpha] = 0 \qquad\qquad [h^\alpha \ell_\alpha] = 0$$

and $[\, | h |^2 \,] = 0$. Our theorem is proved.

b) In an Alfven shock, the tangential velocity of the fluid after the shock is undetermined, but may be determined by the tangential component of the magnetic field according to the following result.

Consider an Alfven shock of type A ; Σ is generated by the integral curves of the field A , with

$$A^\beta = \beta v^\beta + k^\beta$$

One has

$$A^\beta \ell_\beta = \beta u^\beta \ell_\beta + h^\beta \ell_\beta = 0$$

From (VII.3.8) , we deduce

$$\beta^2 (u^\alpha \ell_\alpha)[v^\beta] - (h^\alpha \ell_\alpha)[k^\beta] = 0$$

that is

$$\beta(u^\alpha \ell_\alpha)[\beta v^\beta + k^\beta] = 0$$

The shock being non tangential, one has $[A^\beta] = 0$. We obtain

Theorem. - *In an Alfven shock of* type A (resp B), *the vector* A^β (resp B^β) *is invariant under the shock.*

VIII - HUGONIOT'S FUNCTION AND APPLICATIONS.

VIII.1. - Relativistic Hugoniot's relation for $\ell^{\alpha} \ell_{\alpha} \neq 0$.

We have seen (formula VII.2.8)) that

$$HK = c^2 b^2 / a^2 - L/c^2$$

It follows that, in the set of invariants (VII.3.2) .. (VII.3.6) , the invariant K may be substitued for (VII.3.3), supposing $(h_0^{\alpha} \ell_{\alpha})(h_0^{\alpha} \ell_{\alpha})(h_1^{\alpha} \ell_{\alpha}) \geqslant 0$.

According to (VII.2.6) , the invariance of K may be written

$$\text{(VIII.1.1)} \quad c^2[f^2] - (c^2 a^2 / \ell^{\alpha} \ell_{\alpha})[\tau^2] + 2 \mu [\chi \tau] - (\mu^2 H / c^2)[\chi] = 0$$

But, according to (VII.3.6), one has

$$(\bar{q}_1 - \bar{q}_0) / (\tau_1 - \tau_0) = c^2 a^2 / \ell^{\alpha} \ell_{\alpha}$$

and (VIII.1.1) is equivalent to

$$c^2[f^2] - (\tau_0 + \tau_1)[\bar{q}] + 2 \mu [\chi \tau] - (\mu^2 H / c^2)[\chi] = 0$$

let, if one explicits $\bar{q}$:

$$\text{(VIII.1.2)} \qquad c^2[f^2] - (\tau_0 + \tau_1)[p] - \frac{1}{2} \mu (\tau_0 + \tau_1)[\chi] + 2 \mu [\chi \tau] - \\ -(\mu^2 H / c^2)[\chi] = 0$$

We will substitute for (VIII.1.2) a simpler relation.

a) Study the term $[\chi\tau]$

$$[\chi\tau] = (\chi_1 - \chi_0)\,\tau_1 + \chi_0\,(\tau_1 - \tau_0)$$

or

$$[\chi\tau] = \chi_1\,(\tau_1 - \tau_0) + (\chi_1 - \chi_0)\,\tau_0$$

It follows

$$2[\chi\tau] = (\tau_0 + \tau_1)[\chi] + (\chi_0 + \chi_1)\,[\tau] \tag{VIII.1.3}$$

According to (VIII.1.3), the relation (VIII.1.2) may be written after simplifications

$$c^2[f^2] - (\tau_0 + \tau_1)\,[p] + \frac{1}{2}\,\mu\,(\tau_0 + \tau_1 - (2\,\mu H / c^2)[\chi] + \\ + \mu(\chi_0 + \chi_1)\,[\tau] = 0$$

let

$$c^4[f^2] - c^2(\tau_0 + \tau_1)\,[p] + \frac{1}{2}\,\mu\,(\alpha_0 + \alpha_1)[\chi] + \\ + \mu(\chi_0 + \chi_1)\,[\alpha] = 0 \tag{VIII.1.4}$$

Consider the sum of the two last terms. One has

$$\frac{1}{2}(\alpha_0 + \alpha_1)\,[\chi] + \frac{1}{2}\,(\chi_0 + \chi_1)\,[\alpha] = [\chi\,\alpha]$$

But, according to (VII.3.5)

$$[\chi\alpha] = \chi_1 \alpha_1 - \chi_0 \alpha_0 = \frac{\chi_0 \alpha_0^2}{\alpha_1} - \chi_0 \alpha_0 = - \frac{\chi_0 \alpha_0}{\alpha_1} (\alpha_1 - \alpha_0)$$

The relation (VIII.1.4) may be written

(VIII.1.5)
$$c^2(f_1^2 - f_0^2) - (\tau_0 + \tau_1)(p_1 - p_0) + +(\tau_1 - \tau_0) \frac{1}{2} \mu (\chi_0 + \chi_1 - 2 \frac{\chi_0 \alpha_0}{\alpha_1}) = 0$$

This relation will be called ***Hugoniot's relation*** [9].

b) Given an initial state Y_0 at $x \in \Sigma$, consider the set $\mathcal{E}$ of all the possible states Y satisfying the conditions

(VIII.1.6)
$$H(Y) = H(Y_0) = H \qquad\qquad L(Y) = L(Y_0) = L$$

so that, for $\tau \neq H/c^2$, one has

$$\chi = L/(c^2 \tau - \mu H)^2 = \chi_0 \alpha_0^2 / (c^2\tau - \mu H)^2$$

Among the variables τ, S and $\bar{q}$ one has the functional relation

(VIII.1.7)
$$\bar{q} = p(\tau, S) + \mu L/2 (c^2\tau - \mu H)^2$$

where $p(\tau,S) \geqslant 0$ is obtained by inversion of the equation of state $\tau = \tau(p,S)$ (see Section V.6)). Let P be the half-plane $(\tau, \bar{q})$ defined by $\tau \geqslant 0$. Under the conditions (VIII.1.6), a thermodynamic state of the fluid, defined for example by (τ, p), determines a point $Z = (\tau, \bar{q})$ of P such that

(VIII.1.8) $\overline{q} \geqslant \mu L/2(c^2\tau - \mu H)^2$

This relation defines in P one or two connected and ***convex regions*** $\mathcal{R}$ according as $H \leqslant 0$ or $H > 0$. Conversely, a point $Z = (\tau, \overline{q})$ of $\mathcal{R}$ defines a thermodynamic state (τ, p) of the fluid and values for α and χ. We denote by σ the natural map of $\mathcal{E}$ into P. We are led to introduce ***Hugoniot's function*** $\mathcal{H}(Z_0, Z)$ of relativistic MHD, considered as a function of $Z \in \mathcal{R}$, for a given initial point $Z_0 \in \mathcal{R}$, with [9]

(VIII.1.9) $$\mathcal{H}(Z_0, Z) = c^2(f^2 - f_0^2) - (\tau + \tau_0)(p - p_0) + $$

$$+ (\tau - \tau_0)(1/2)\mu(\chi + \chi_0 - 2\chi_0 \alpha_0/\alpha)$$

It is clear that $\mathcal{H}(Z_0, Z_0) = 0$ and that (VIII.1.5) may be writen $\mathcal{H}(Z_0, Z_1) = 0$ for $\sigma(Y_1) = Z_1$. It is a detailed study of the behavior of Hugoniot's function that furnishes a good part of the results that we seek.

VIII.2. - Differential of Hugoniot's function.

By differentiation of (VIII.1.9) under the conditions (VIII.1.6), one obtains

$$d\mathcal{H} = 2c^2 f df - (\tau + \tau_0)\, dp - (p - p_0) d\tau + (1/2)\mu(\chi + \chi_0 - 2\chi_0 \alpha_0/\alpha)\, d\tau$$

$$+ (1/2)\mu(\tau - \tau_0)(d\chi + 2(\chi_0 \alpha_0/\alpha^2)\, d\alpha)$$

But one has

$$c^2 f df = f\,\Theta\, dS + \tau\, dp \qquad\qquad d\alpha = c^2\, d\tau$$

It follows

$$d\mathcal{H} = 2f\,\Theta\,dS + (\tau - \tau_0)\,dp - (p - p_0)d\tau + (1/2)\,\mu\,(\chi + \chi_0)$$

$$-2(\chi_0\,\alpha_0/\alpha)\,d\tau + (1/2)\mu(\tau - \tau_0)\,(d\chi + 2c^2(\chi_0\,\alpha_0/\alpha^2)\,d\tau)$$

One has

$$dp = d\bar{q} - (1/2)\,\mu\,d\chi \qquad (p - p_0) = (\bar{q} - \bar{q}_0) - (1/2)\,\mu(\chi - \chi_0)$$

One obtains

$$d\mathcal{H} = 2f\,\Theta\,dS + (\tau - \tau_0)\,d\bar{q} - (\bar{q} - \bar{q}_0)d\tau +$$

$$+ \mu\,(\chi - (\chi_0\,\alpha_0/\alpha)\,d\tau + \mu(\alpha - \alpha_0)\,(\chi_0\,\alpha_0/\alpha^2)\,d\tau$$

let, after simplifications :

$$d\mathcal{H} = 2f\,\Theta\,dS + (\tau - \tau_0)\,d\bar{q} - (\bar{q} - \bar{q}_0)d\tau + \mu\,(\chi - (\chi_0\,\alpha_0^2/\alpha^2)\,d\tau$$

But, according to (VIII.1.6) $\chi\,\alpha^2 = \chi_0\,\alpha_0^2$. One has

(VIII.2.1) $$d\mathcal{H} = 2f\,\Theta\,dS + (\tau - \tau_0)\,d\bar{q} - (\bar{q} - \bar{q}_0)d\tau$$

In the plane $(\tau\,,\,\bar{q})$ introduce as parameter the slope m of the straight line $(Z_0\,,\,Z)$ of $\mathcal{R}$

(VIII.2.2) $$(\bar{q} - \bar{q}_0) = m(\tau - \tau_0)$$

By differentiation, one has

$$d\bar{q} = m\,d\tau + (\tau - \tau_0)\,dm$$

One obtains

$$(\tau - \tau_0)\,d\bar{q} - (\bar{q} - \bar{q}_0)\,d\tau = (\tau - \tau_0)^2\,dm$$

and (VIII.2.1) may be written

(VIII.2.3) $$d\mathscr{H} = 2f\,\Theta\,dS + (\tau - \tau_0)^2\,dm$$

VIII.3. - Differential of S along a straight line ($\ell^\alpha \ell_\alpha \neq 0$)

a) We study the differential of S along the part in $\mathscr{R}$ of a straight line Δ of P with slope m. One has

$$\tau'_S\,dS = d\tau - \tau'_p\,dp$$

and

$$dp + (1/2)\,\mu\,d\chi = d\bar{q} = m\,d\tau$$

One obtains

$$\tau'_s\,dS = (1 - m\tau'_p)\,d\tau + (1/2)\mu\,\tau'_p\,d\chi$$

By differentiation of $\chi\,\alpha^2 = \text{const.}$, one has

$$\alpha\,d\chi + 2\,c^2\,\chi\,d\tau = 0$$

It follows

$$\text{(VIII.3.1)} \qquad \tau'_s \, dS = \tau'_p \, (\tau'^{-1}_p - c^2 \mu \chi \alpha^{-1} - m) \, d\tau$$

Given a point Z of $\mathcal{R}$, let $\mathcal{S}$ be ***the isentropic curve*** of Z corresponding to the value S of the specific entropy, given by (VIII.1.7). The slope of $\mathcal{S}$ at Z is precisely, according to the definition of $\bar{q}$ and (V.6.3) :

$$(d\bar{q} / d\tau)_{\mathcal{S}} = \tau'^{-1}_p - c^2 \mu \chi \alpha^{-1}$$

We set for $\mathcal{S}$ and for a straight line Δ with slope m passing through Z

$$\text{(VIII.3.2)} \qquad Q(m , Z) = (d\bar{q} / d\tau)_{\mathcal{S}} (Z) - m$$

and one has :

$$\text{(VIII.3.3)} \qquad \tau'_s \, dS = \tau'_p \, Q(m) \, d\tau$$

b) Study $P(\ell)$ for a state Y. One has, according to the definition of H ,

$$(h^\alpha \ell_\alpha)^2 = a^2 (H + |h|^2 / r^2)$$

It follows from (VI.6.15) :

$$\text{(VIII.3.4)} \qquad P(\ell) = c^2 rf (\gamma - 1)(u^\alpha \ell_\alpha)^4 +$$

$$+ (c^2 rf + \mu |h|^2 (\gamma - 1))(u^\alpha \ell_\alpha)^2 \ell^\beta \ell_\beta - \mu a^2 H \ell^\beta \ell_\beta$$

One obtains

(VIII.3.5) $$P(\ell) = c^2 rf(\gamma - 1)(u^\alpha \ell_\alpha)^4 +$$

$$+ \{(c^2 rf + \mu(\chi - (a^2/\ell^\alpha \ell_\alpha)H(\gamma-1)\}(u^\alpha \ell_\alpha)^2 \ell^\beta \ell_\beta - \mu a^2 H \ell^\beta \ell_\beta$$

We have seen that a point Z of $\mathcal{R}$ defines values for the thermodynamic variables, for α and χ. We introduce now the condition

(VIII.3.6) $$r u^\alpha \ell_\alpha = r_0 u_0^\alpha \ell_\alpha = a$$

If such is the case, the right-hand side of (VIII.3.5) defines for each point Z of $\mathcal{R}$ a number $P(\ell)(Z)$.

c) According to (VII.3.6) consider the points Z of the straight line Δ_a issuing from Z_0 with slope $m_a = c^2 a^2 / \ell^\alpha \ell_\alpha$. One has

$$Q(m_a) = \tau_p'^{-1} - c^2 \mu \chi \alpha^{-1} - c^2 a^2 / \ell^\alpha \ell_\alpha$$

According to (VIII.3.5), one obtains

$$P(\ell)/a^2 = ((\gamma-1)/r^2)(c^2 a^2 \tau - \mu a^2 H + \mu \chi (\ell^\beta \ell_\beta)) + (c^2 \tau - \mu H) \ell^\beta \ell_\beta$$

Where $c^2 \tau_p' = -(\gamma-1)/r^2$. It follows

$$P(\ell)/a^2(\ell^\alpha \ell_\alpha) = \alpha - c^2 \tau_p' \left(\frac{a^2}{\ell^\alpha \ell_\alpha} \alpha + \mu \chi \right)$$

let

$$P(\ell) / a^2 \tau_p' \alpha (\ell^\alpha \ell_\alpha) = \tau_p'^{-1} - c^2 a^2 / (\ell^\alpha \ell_\alpha) - c^2 \mu \chi \alpha^{-1}$$

One obtains

$$Q(m_a) = P(\ell) \,/\, a^2 c'_p \alpha (\ell^\alpha \ell_\alpha) \tag{VIII.3.7}$$

It follows from (VIII.3.3)

$$\tau'_s \alpha \, dS = (P(\ell)/c^2 a^4) \, d\bar{q} \tag{VIII.3.8}$$

VIII.4. - The case where $\ell^\alpha \ell_\alpha = 0$ is impossible.

a) Let $Y_0 \to Y_1$ be a shock such that the wave front Σ satisfies $\ell^\alpha \ell_\alpha = 0$. According to (VII.2.2), one has

$$W^\beta \ell_\beta = a^2 \alpha$$

and in the shock $[\alpha] = 0$ and thus $[\tau] = 0$.

Study

$$W^\beta W_\beta / a^2 = (c^2\tau + \mu |h|^2/r^2)^2 r^2 - \mu^2 ((h^\alpha \ell_\alpha)^2/a^2) |h|^2 -$$

$$- 2q (c^2\tau + \mu |h|^2/r^2) + 2q \mu (h^\alpha \ell_\alpha) / a^2$$

where $q = p + (1/2) \mu |h|^2$. One obtains after simplifications

$$W^\beta W_\beta / c^2 a^2 = c^2 f^2 - 2\tau p + \tau \cdot \mu |h|^2 + 2 (\mu H/c^2) p$$

We set, since $\tau = \tau_0$,

(VIII.4.1) $$\mathcal{K}(\tau_0 , p_0 ; \tau_0 , p) = c^2 (f^2 - f_0^2) - 2 \tau_0 (p-p_0) +$$

$$+ \tau_0 (\mu | h |^2 - \mu | h_0 |^2) + 2(\mu H/c^2)(p-p_0)$$

$\mathcal{K}$ plays here the role of Hugoniot's function. In the shock, we have

$$\mathcal{K}(\tau_0 , p_0 ; \tau_0 , p_1) = 0$$

b) Consider in the plane (τ , p), the tehrmodynamic states satisfying

(VIII.4.2) $$H(Y) = H(Y_0) = H \qquad b(Y)/a(Y) = b(Y_0)/a(Y_0) = b/a$$

where H is $\leqslant 0$ according to Lemma 2 (part 3) of Section (VII.2). Introduce the straight line Δ issuing of (τ_0 , p_0) satisfying (VIII.4.2) and such that $\tau = \tau_0$. One has, according to the definition of H

$$\frac{f^2(h^\alpha \ell_\alpha)^2}{a^2} - | h |^2 \tau_0^2 = Hf^2$$

or

(VIII.4.3) $$b^2 / a^2 - \tau_0^2 | h |^2 = Hf^2$$

By differentiation one has

(VIII.4.4) $$c^2 \tau_0^2 d | h |^2 = - 2H (f \Theta dS + \tau_0 dp)$$

It follows that the differential of $\mathcal{K}$ along Δ gives

$$d\mathcal{K} = 2f \Theta dS + \tau_0 \mu d | h |^2 + 2 (\mu H/c^2) dp$$

or, according to (VIII.4.4)

$$d\mathcal{K} = 2f\,\Theta\,dS - 2(\mu H/c^2\,\tau_0)\,(f\,\Theta\,dS + \tau_0\,dp) + 2\,(\mu H/c^2)\,dp$$

let

$$d\mathcal{K} = 2(1 - \mu H / c^2\,\tau_0)\,f\,\Theta\,dS \quad (H \leqslant 0) \tag{VIII.4.5}$$

τ being constant, one has along Δ

$$\tau'_s\,dS = -\,\tau'_p\,dp$$

Under the hypotheses $\tau'_p < 0$, $\tau'_s \neq 0$, one has $(dS/dp)_\Delta \neq 0$ along Δ . But one has

$$\mathcal{K}\,(\tau_0\,,p_0\,;\tau_0\,,p_0) = \mathcal{K}\,(\tau_0\,,p_0\,;\tau_0\,,p_1) = 0$$

and the function $\mathcal{K}\,(\tau_0\,,p_0\,;\tau_0\,,p)$ is stationary at at least one point of the segment joining the states before and after the shock. It is the same for S according to (VIII.4.5) and we arrive at a contradiction. One has

Proposition. *Under the hypotheses* $\tau'_p < 0$, $\tau'_s \neq 0$, *any front* Σ *of a magnetodynamic shock wave never satisfies* $\ell^\alpha\,\ell_\alpha > 0$.

Under the hypotheses (H_1) and (H_2) of Section (V.6), this circonstance is forbidden.

VIII.5. - Orientation of shock wave fronts in the time.

a) Consider a shock wave that is not a zero shock and not an Alfven shock. If Σ is the front of the shock wave suppose that $\ell^\alpha \ell_\alpha > 0$. At the point x of Σ one has the shock $Y_0 \to Y_1$ and we know that $H \leqslant 0$ and $\alpha > 0$ again. Set $Z_0 = \sigma(Y_0)$, $Z_1 = \sigma(Y_1)$ in the half-plane P of the points $(\tau , \overline{q})$; we denote by m the slope of the segment (Z_0 , Z_1) .

$\ell^\alpha \ell_\alpha$ being > 0 , we deduce from (VIII.4.4) that $P(\ell)$ is > 0 under the hypothesis $\tau'_p < 0$ (or $\gamma > 1$). If $\tau'_S \neq 0$, it follows from (VIII.3.8) that one has $(dS/d\overline{q}) \neq 0$ along (Z_0 , Z_1) . But, according to (VIII.2.3), one has over the segment (Z_0 , Z_1)

(VIII.5.1) $$d\mathcal{H} = 2 f \Theta \, dS$$

and

$$\mathcal{H}(Z_0 , Z_0) = \mathcal{H}(Z_0 , Z_0) = 0$$

The function $\mathcal{H}(Z_0 , Z)$ is stationary at at least one point of the segment (Z_0 , Z_1). It is the same according to (VIII.5.1) and we arrive at a contradiction.

Section (VIII.4) and Section (VIII.5) show that, under our hypotheses, $\ell^\alpha \ell_\alpha$ is < 0 and Σ is oriented *in the time.* We have

Theorem. - *Under the hypotheses $\tau'_p < 0$, $\tau'_s \neq 0$, any front of a magnetohydrodynamic shock wave is oriented in time. If v_0^Σ and v_1^Σ are the speeds of Σ with respect to the fluid before and after the shock, one has*

$$v_0^{\Sigma} < c \qquad\qquad v_1^{\Sigma} < c$$

b) From Lemma 2 of Section (VII.2), we know that $\chi = \psi^2$ is positive. With a suitable choice of the sign of ψ, we may substitute for the second condition (VIII.1.6) the condition

(VIII.5.2) $$\psi\alpha = \psi_0\,\alpha_0$$

One has then

$$\chi + \chi_0 - 2\,\chi_0\,\alpha_0\,/\,\alpha = \psi^2 + \psi_0^2 - 2\,\psi\,\psi_0 = (\psi - \psi_0)^2$$

Hugoniot's function (VIII.1.9) may be written :

(VIII.5.3) $$\mathcal{H}(Z_0\,,\,Z) = c^2\,(f^2 - f_0^2) - (\tau + \tau_0)(p - p_0) + $$
$$+ (\tau - \tau_0)\,\frac{1}{2}\,\mu\,(\psi - \psi_0)^2$$

and Hugoniot's relation is

(VIII.5.4) $$\mathcal{H}(Z_0\,,\,Z_1) = c^2\,(f_1^2 - f_0^2) - (\tau_1 + \tau_0)(p_1 - p_0)$$
$$+ (\tau_1 - \tau_0)\,\frac{1}{2}\,\mu(\psi_1 - \psi_0)^2$$

VIII.6. - Thermodynamic inequalities for the shocks.

We recall that we have the following hypotheses on the state function $\tau(p,S)$ (see Section IV.6))

(H_1) *One has* $\tau'_p < 0$ *and* $\tau''_{p^2} > 0$.

(H_2) τ'_S *is* $\neq 0$ *and there exists a function* $S = S(p, \tau)$ *whose unique inverse is* $\tau = \tau(p,S)$.

We have in variables (p,S) (see Section (V.1))

(VIII.6.1) $$c^2 f'_p = V > 0 \qquad\qquad c^2 f'_S = \Theta > 0$$

It follows

(VIII.6.2) $$c^2 (f^2)'_p = 2\tau$$

Consider a shock $Y_0 \to Y_1$ that is not a zero shock neither an Alfven shock ; $Z_0 = \sigma(Y_0)$ and $Z_1 = \sigma(Y_1)$ are connected by Hugoniot's relation (VIII.5.4), which is symmetric in 0 and 1 . According to (VII.3.1), we suppose that at any point $x \in \Sigma$ one has $S_1 \geqslant S_0$.

a) ***Suppose*** $S_1 = S_0$. If $p_1 \neq p_0$, one may suppose for example that $p_1 > p_0$. One then has $\tau_1 < \tau_0$ since $\tau'_p < 0$. From (VIII.6.2) one deduces

$$c^2(f^2(p_1, S_0) - f^2(p_0, S_0)) = 2\int_{p_0}^{p_1} \tau(p, S_0)\, dp$$

From the convexity condition $\tau''_{p^2} > 0$, it follows

$$c^2(f_1^2 - f_0^2) < (\tau(p_0, S_0) + \tau(p_1, S_0))(p_1 - p_0)$$

let

$$c^2(f_1^2 - f_0^2) - (\tau_0 + \tau_1)(p_1 - p_0) < 0$$

Hugoniot's relation (VIII.5.4) gives then $\tau_1 > \tau_0$, that is a contradiction. One thus has $p_1 = p_0$ and our shock is either zero or an Alfven shock.

Proposition 1. - *Under the hypotheses* (H_1), (H_2) *a non zero shock that is not an Alfven shock, satisfies necessarily* ([9] and [71])

$$S_1 > S_0$$

b) Consider such a shock and suppose $p_1 \geqslant p_0$. One has

$$c^2(f^2(p_1, S_1) - f^2(p_0, S_1)) = 2\int_{p_0}^{p_1} \tau(p, S_1)\, dp \geqslant 2\tau_1(p_1 - p_0)$$

S_1 being $> S_0$ and f'_S being > 0 , one has as a consequence

$$c^2(f_1^2 - f_0^2) - 2\tau_1(p_1 - p_0) > 0$$

From Hugoniot's relation, one deduces

$$(\tau_1 - \tau_0)\{(p_1 - p_0) + (1/2)\mu(\psi_1 - \psi_0)^2\} < 0$$

and one has $\tau_1 < \tau_0$. Thus $p_1 \geqslant p_0$ implies $\tau_1 < \tau_0$.

Proposition 2. *For a non zero shock that is not an Alfven shock, the inequality* $p_1 \geqslant p_0$ *implies, under the hypotheses* (H_1), (H_2), *that* $\tau_1 < \tau_0$.

c) ***Suppose*** $\tau'_S < 0$. For the considered shock, one has $S_1 > S_0$: Suppose that $p_1 < p_0$. From the convexity condition $\tau''_{p^2} > 0$, one deduces

$$c^2\{f^2(p_0, S_1) - f^2(p_1, S_1)\} = 2\int_{p_1}^{p_0} \tau(p, S_1)\,dp \leqslant \{\tau(p_0, S_1) + \tau(p_1, S_1)\}(p_0 - p_1)$$

It follows

$$c^2(f_0^2 - f_1^2) - (\tau_0 + \tau_1)(p_0 - p_1) < 0$$

let

$$c^2(f_1^2 - f_0^2) - (\tau_0 + \tau_1)(p_1 - p_0) > 0$$

From Hugoniot's relation, one deduces $\tau_1 < \tau_0$.

Proposition 3. *For* $\tau'_S < 0$, *if for a non zero shock that is not an Alfven shock, one has* $p_1 < p_0$, *one has necessarily, under the hypotheses* (H_1), (H_2), $\tau_1 < \tau_0$.

d) ***Suppose*** $\tau'_S > 0$. For the considered shock, one has $S_1 > S_0$. Suppose that $p_1 < p_0$. From the convexity condition, one deduces

$$c^2\{f^2(p_0, S_0) - f^2(p_1, S_0)\} = 2\int_{p_1}^{p_0} \tau(p, S_0)\,dp \leqslant \{\tau(p_0, S_0) + \tau(p_1, S_0)\}(p_0 - p_1)$$

It follows

$$c^2(f_0^2 - f_1^2) - (\tau_0 + \tau_1)(p_0 - p_1) < 0$$

let

$$c^2(f_1^2 - f_0^2) - (\tau_0 + \tau_1)(p_1 - p_0) > 0$$

From Hugoniot's relation, one deduces $\tau_1 < \tau_0$ that is a contradiction with $\tau_p' < 0$, $\tau_S' > 0$. The hypothesis $p_1 < p_0$ is false.

Proposition 4. *For* $\tau_S' > 0$, *one has for a non zero shock that is not an Alfven shock and under the hypotheses* (H_1) , (H_2) , $p_1 > p_0$ *and thus* $\tau_1 < \tau_0$.

e) These different results may be assembled in the following statement ([9] and [11]).

Theorem 1. *For a non zero shock that is not an Alfven shock , one has under the hypotheses* (H_1) , (H_2) :

(VIII.6.3) $\quad S_1 > S_0 \qquad \tau_1 < \tau_0$

2) *Moreover, if* $\tau_S' > 0$, *one has*

(VIII.6.4) $\quad p_1 > p_0 \qquad f_1 > f_0 \qquad V_1 < V_0$

(compression shock)

One deduces from this theorem that $\alpha_1 < \alpha_0$. It may be shown (Ann. 1) that a shock wave that is not an Alfven shock is compatible with ordinary Alfven waves ***only if*** $\alpha_1 \alpha_0 > 0$. Thus there are two types of shocks [9]:

slow shocks for which $\alpha_1 < \alpha_0 < 0$

rapid shocks for which $0 < \alpha_1 < \alpha_0$

VIII.7. - Isentropic curves and Hugoniot's curve.

a) An initial state Y_0 being given, we consider the set $\mathcal{E}$ of the states Y satisfying

(VIII.7.1) $$H(Y) = H(Y_0) = H$$

$$(\psi\alpha)(Y) = \psi_0 \alpha_0 \quad (\alpha \alpha_0 > 0)$$

with $(h^\alpha \ell_\alpha)(h_0^\alpha \ell_\alpha) \geqslant 0$; If $Z = \sigma(Y)$, for $H > 0$ the straight line $\tau = \mu H/c^2$ of the half plane P defined by $(\tau, \bar{q})$ is forbidden. In P , the isentropic curve $\mathcal{S}$ corresponding to the value S of the entropy is defined by

(VIII.7.2) $$\bar{q} = p(\tau, S) + \frac{\mu \psi_0^2 \alpha_0^2}{2(c^2\tau - \mu H)^2} = F(\tau, S)$$

By inversion of $\tau = \tau(p, S)$ in $p = p(\tau, S)$, one has (see Section (V.6)), one has

(VIII.7.3) $\qquad p'_\tau = \tau'^{-1}_p \qquad p'_\tau = -\,\tau'^{-1}_p\,\tau'_s \qquad p''_{\tau 2} = -\,\tau'^{-3}_p\,\tau''_{p2}$

By differentiation of (VIII.7.2) along $\mathcal{S}$, one obtains

$$(d\bar{q}/d\tau^2\,)_{\mathcal{S}} = \tau'^{-1}_p - c^2\,\mu\,\psi_0^2\,\alpha_0^2\,/\,\alpha^3$$

and by a second differentiation

$$(d^2\bar{q}/d\tau^2\,)_{\mathcal{S}} = -\tau'^{-3}_p\;\tau''_{p2} + 3\,c^4\,\mu\,\psi_0^2\,\alpha_0^2\,/\,\alpha^4$$

We set :

(VIII.7.4) $\qquad M = \tau''_{p2} - (3\,c^4\,\mu\,\tau'^{3}_p\,\psi_0^2\,\alpha_0^2\,)\,/\,\alpha^4 > 0$

Under the hypothesis (H_i) M is ***strictly positive*** and one has

(VIII.7.5) $\qquad (d^2\,\bar{q}\,/\,d\tau^2\,)_{\mathcal{S}} = -\,\tau'^{-3}_p\,M$

Thus ***each isentropic curve*** $\mathcal{S}$ ***is strictly concave.***

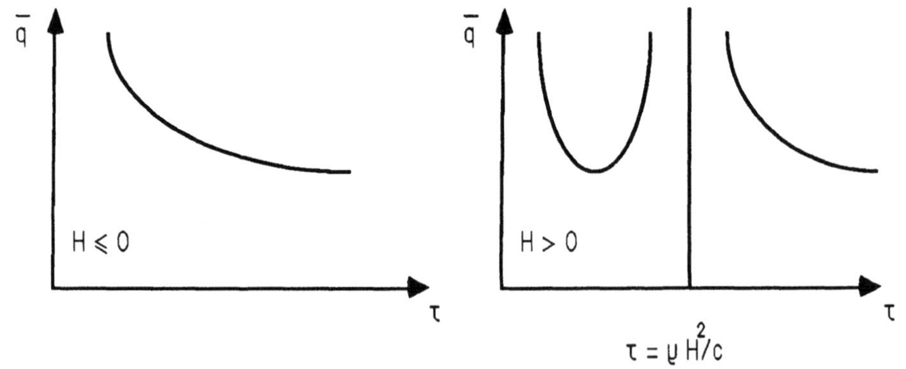

We note that the function F defined by (VIII.7.2) satisfies

(VIII.7.6) $$F'_S = -\tau'^{-1}_p \tau'_S \qquad F^{\cdot}_{\tau^2} = -\tau'^{-3}_p M$$

b) Let Δ be a straight line of slope m in the half-plane P . One has in a region $\mathcal{R}$.

$$d\overline{q} = m\, d\tau$$

We will prove.

Lemma. *In a point* Z_s *of* Δ *where* S *is stationary, one has under the hypotheses* (H_1), (H_2) .

(VIII.7.7) $$\{\tau'_S (d^2S/d\tau^2)_\Delta\} (Z_s) = -(\tau'^{-2}_p M)(Z_1) < 0$$

In fact, by differentiation along Δ with respect to τ of the functional relation

$$\overline{q} = F(\tau , S)$$

one has

$$F'_\tau + F'_S (dS/d\tau)_\Delta = m$$

By a second differentiation, one obtains

$$F''_{\tau^2} + 2F''_{S\tau} (dS/d\tau)_\Delta + F''_{S^2}(dS/d\tau)^2_\Delta + F'_S(d^2S/d\tau^2)_\Delta = 0$$

In a point Z_s of Δ where $(dS/d\tau)_\Delta$ is null, one has

$$(F''_{\tau^2})_{Z_S} + (F'_S)_{Z_S} (d^2S/d\tau^2)_\Delta (Z_S) = 0$$

It follows from (VIII7.6)

$$(d^2S/d\tau^2)_\Delta (Z_S) = -\{ \tau'_p \tau'^{-3}_p M \tau'^{-1}_S \}(Z_1) = -(\tau'^{-1}_S \tau'^{-2}_p M)(Z_S)$$

and our lemma is proved.

c) We consider in P only the region defined in $\mathcal{R}$ by $\tau \leqslant \tau_0$. The ***Hugoniot curve*** $\mathcal{H}$ is the set of points Z of this region such that $\mathcal{H}(Z_0, Z) = 0$.

Let $Z_1 \neq Z_0$ be an arbitrary point of $\mathcal{H}$ and Δ the straight line (Z_0, Z_1) with slope m ; $\mathcal{H}(Z_0, Z)$ is stationary at at least one point Z_S of the segment $(Z_0\ Z_1)$. From (VIII.2.3) S is stationary in Z_S and it follows from the lemma of b that Z_S is unique and corresponds, according to the sign of τ'_S, either to a strict maximum or to a strict minimum of S over Δ. One thus has at Z_1

(VIII.7.8) $$(\tau'_S dS/d\tau)_\Delta)(Z_1) > 0$$

let, according to (VIII.3.3), $Q(m, Z_1) < 0$. In particular $Q(m, Z_1)$ does not vanish over $\mathcal{H}$. On the other hand

(VII.7.9) $$(\tau'_S(dS/d\tau)_\Delta (Z_0) < 0$$

that is $Q(m ; Z_0) > 0$. If m_0 is the slope in Z_0 of the isentropic curve $\mathcal{S}_0$ of this point, one has

(VIII.7.10) $m < m_0$ $((d\bar{q}/d\tau)_{\mathcal{S}_0} - m)(Z_0) > 0$

d) Suppose that there is $Z_1 \in \mathcal{H}$ such that $(dS/d\tau)_{\mathcal{H}}(Z_1) = 0$. The isentropic curve of Z_1 is tangent at Z_1 to $\mathcal{H}$. It follows from (VIII.2.3) and (VIII.3.2) that the straight-line $\Delta = (Z_0, Z_1)$ with slope m is also tangent at Z_1 to these curves and that $Q(m, Z_1) = 0$, which is impossible. Thus, S ***is a strictly monotone function of*** τ ***over each connected component of*** $\mathcal{H}$.

One has

$$(dS/dp)_{\mathcal{H}} = (dS/d\tau)_{\mathcal{H}}(\tau'_p + \tau'_S(dS/dp)_{\mathcal{H}})$$

and $(dS/dp)_{\mathcal{H}}$ is never zero ; S ***is also a strictly monotone function of*** p ***over each connected component of*** $\mathcal{H}$.

VIII.8. - Speeds of shock waves.

a) Let $Y_0 \to Y_1$ be a non zero shock that is not an Alfven shock and Δ_a the corresponding straight line (Z_0, Z_1) of P with slope m_a. According to (VIII.7.8), (VIII.7.9), one has

(VIII.8.1) $(\tau'_S(dS/d\tau)_{\Delta_a})(Z_0) < 0$ $(\tau'_S(dS/d\tau)_{\Delta_a})(Z_1) > 0$

It follows from (VIII.2.3) and (VIII.3.7)

(VIII.8.2) $\alpha_0 P(\ell)(Z_0) > 0$ $\alpha_1 P(\ell)(Z_1) < 0$

Geometrically, (VIII.8.1) has the following interpretation

$$(\text{VIII.8.3}) \qquad m_0 = (d\bar{q}/d\tau)_{\mathcal{S}0}(Z_0) > m_a \qquad m_1 = (d\bar{q}/d\tau)_{\mathcal{S}1}(Z_1) < m_a$$

b) Inequalities (VIII.8.2) may be interpreted in terms of speeds of the shock wave front Σ and of magnetosonic and Alfven speeds before and after the shock. In fact, one has $\alpha = D(\ell)/a^2$ and the relation (VI.7.6)

$$D(\ell) = (c^2 rf + \mu |h|^2)((u^\alpha \ell_\alpha)^2 - \ell^\alpha \ell_\alpha)(y^\Sigma - \mu h_n^2 / (c^2 rf + \mu |h|^2)$$

may be written

$$(\text{VIII.8.4}) \quad \frac{a^2}{(c^2 rf + \mu |h|^2)(u^\alpha \ell_\alpha)^2 - \ell^\alpha \ell_\alpha)} \qquad \alpha = \frac{(v^\Sigma)^2}{c^2} - \frac{(v_A)^2}{c^2}$$

consider a rapid shock $(\alpha_1 < \alpha_0 < 0)$. According to (VIII.8.4), one has

$$v_0^\Sigma > v_0^A \qquad\qquad v_1^\Sigma > v_1^A$$

From (VIII.8.2) it follows that $P(\ell)(Z_0)$ is > 0 and $P(\ell)(Z_1)$ is < 0 . According to (VI.7.3) one has $\Pi((v_0^\Sigma)^2 / c^2) > 0$ and $\Pi((v_1^\Sigma)^2 / c^2) < 0$. One obtains

$$v_0^\Sigma > v_0^{RM} \qquad\qquad v_1^\Sigma < v_1^{RM}$$

Consider now a slow shock $(\alpha_1 < \alpha_0 < 0)$. According to (VIII.8.4), one has

$$v_0^\Sigma < v_0^A \qquad\qquad v_1^\Sigma < v_1^A$$

From (VIII.8.2) it follows that $P(\ell)(Z_0)$ is < 0 and $P(\ell)(Z_1)$ is > 0 . In the same way one obtains

$$v_0^{\Sigma} > v_0^{SM} \qquad\qquad v_1^{\Sigma} < v_1^{SM}$$

One has ([9] and [11])

Theorem. *Under the hypotheses* (H_1), (H_2), *the speeds* v_0^{Σ} *and* v_1^{Σ} *of the shock wave front* Σ *satisfy the inequalities*

1) *for a rapid shock*

$$v_1^{SM} < v_0^{A} < v_0^{RM} < v_0^{\Sigma} \qquad\qquad v_1^{SM} < v_1^{A} < v_1^{\Sigma} < v_1^{RM}$$

2) *for a slow shock*

$$v_0^{SM} < v_0^{\Sigma} < v_0^{A} < v_0^{RM} \qquad\qquad v_1^{\Sigma} < v_1^{MS} < v_1^{A} < v_1^{RM}$$

VIII.9. - Isentropic curves and weak shocks.

a) Consider the region $\tau \leqslant \tau_0$ of $\mathcal{R}$ $(\alpha\alpha_0 > 0)$ the isentropic curve $\mathcal{S}_0$ corresponding to $S = S_0$. According to (VIII.2.1), one has along $\mathcal{S}_0$

$$(d\mathcal{H}/d\tau)_{\mathcal{S}_0} = (\tau - \tau_0)(d\bar{q}/d\tau)_{\mathcal{S}_0} - (\bar{q} - \bar{q}_0)$$

By differentiation and according to (VIII.7.5) one obtains

$$(d^2\mathcal{H}/d\tau^2)_{\mathcal{S}_0} = (\tau - \tau_0)(d^2\bar{q}/d\tau^2)_{\mathcal{S}_0} = -(\tau - \tau_0)\,\tau_p'^{-3}\,M < 0$$

One sees that over the considered arc of $\mathcal{S}_0$

(VIII.9.1) $\qquad (d\mathcal{H}/d\tau)_{\mathcal{S}_0} > 0 \qquad \mathcal{H}(Z_0, Z) < 0 \qquad$ for $Z \in \mathcal{S}_0$ $(Z \neq Z_0)$

$\mathcal{S}_0$ does not meet $\mathcal{H}$ at a point different from Z_0

b) We shall establish in the following that $\mathcal{H}$ is connected. Consider for the moment $\mathcal{H}$ in the neighborhood of Z_0 . One has according to (VIII.2.1)

$$2f\,\Theta\,(dS/d\tau)_{\mathcal{H}} + (\tau - \tau_0)\,(d\bar{q}/d\tau)_{\mathcal{H}} - (\bar{q} - \bar{q}_0) = 0 \tag{VIII.9.2}$$

Thus $(dS/d\tau)_{\mathcal{H}}\,(Z_0) = 0$. By differentiation of (VIII.9.2) one obtains

$$2f\,\Theta\,(d^2S/d\tau^2)_{\mathcal{H}} + 2\,(d(f\,\Theta)/d\tau)_{\mathcal{H}}\,(dS/d\tau)_{\mathcal{H}} + \tag{VIII.9.3}$$

$$+ (\tau - \tau_0)(d^2\bar{q}/d\tau^2)_{\mathcal{H}} = 0$$

and one has also $(d^2S/d\tau^2)_{\mathcal{H}}\,(Z_0) = 0$. Thus Hugoniot's curve $\mathcal{H}$ and the isentropic curve Z_0 have contact of the second order at Z_0 . By differentiation of (VIII.9.3) one has at Z_0 :

$$\{2f\,\Theta\,(d^3S/d\tau^3)_{\mathcal{H}} + (d^2\bar{q}/d\tau^2)_{\mathcal{H}}\}\,(Z_0) = 0$$

One has at Z_0 according to (VIII.7.5)

$$(d^2\bar{q}/d\tau^2)_{\mathcal{H}}\,(Z_0) = (\ell^2\bar{q}/d\tau^2)_{\mathcal{S}_0}\,(Z_0) = -\,(\tau_p'^{3}\,M)\,(Z_0) \tag{VIII.9.4}$$

It follows from (VIII.9.4)

$$(d^3S/d\tau^3)_{\mathcal{H}} = \{\frac{1}{2f\,\Theta}\,\tau_p'^{-3}\,M\}\,(Z_0) \tag{VIII.9.5}$$

According to the contact between $\mathscr{H}$ and $\mathscr{S}_0$, one has at Z_0

$$(d\tau / dp)_{\mathscr{H}} (Z_0) = (d\tau / dp)_{\mathscr{S}_0} (Z_0) = \tau'_p (Z_0)$$

and thus at this point :

$$(d^3 S / dp^3)_{\mathscr{H}} (Z_0) = (d^3 S / d\tau^3)_{\mathscr{H}} (Z_0)\, \tau'^3_p (Z_0)$$

One obtains thus at Z_0 :

(VIII.9.6) $$(d^3 S / dp^3)_{\mathscr{H}} (Z_0) = \{ M / 2 f \Theta \} (Z_0)$$

and in the neighborhood of Z_0 , one has along $\mathscr{H}$

(VIII.9.7) $$(S - S_0) = \{ (1/12 f \Theta) M \} (Z_0) (p - p_0)^3$$

One sees that *in a weak shock, the increase* $(S_1 - S_0)$ *of entropy is of third order with respect to the strenght* $(p_1 - p_0)$ *of the shock.*

VIII.10. - The case $\tau'_S > 0$ [9].

We suppose that we have the hypotheses (H_1) $(\tau'_p < 0 , \tau''_{p^2} > 0)$ and (H_2) *with* $\tau'_S > 0$. Moreover

$$p = p(\tau , S) \to \infty \quad \text{when} \quad \tau \to 0$$

We see that $\overline{q}$ may be arbitrarily large.

We will show that if $\alpha_0 P(\ell) Z_0$ is > 0 , the equations of shock (VII.3.2) – (VII.3.6) have a unique non trivial solution with $\alpha_0 \alpha_1 > 0$.

a) We denote by m_0 the slope at Z_0 of $\mathcal{S}_0$ ($\tau \leqslant \tau_0$) or of $\mathcal{H}$ (see (VIII.9)). Let Δ be a straight line issued from Z_0 with slope m such that

(VIII.10.1) $\qquad m < m_0$

$\mathcal{S}_0$ being strictly concave, Δ meets $\mathcal{S}_0$ at a unique point $Z_A \neq Z_0$. According to (VIII.9.1) one has

(VIII.10.2) $\qquad \mathcal{H}(\mathcal{H}_0 , Z_A) < 0$

But $S(Z_0) = S(Z_A) = S_0$ and S is stationary at a unique point Z_S of the segment (Z_0 , Z_A) (see (VIII.7) c) corresponding to a strict maximum of S over Δ . According to (VIII.7.9) one has $(dS/d\tau)_\Delta$ $(\tau_0) < 0$ and s0$(d\mathcal{H} / d\tau)_\Delta (Z_0) < 0$. When Z_0 goes from Z_0 to Z_A , $\mathcal{H}(Z_0 , Z)$ is first positive. According to (VIII.2.3) $\mathcal{H}(Z_0 , Z)$ has the same behavior as $S(Z)$ over Δ . It follows from (VIII.10.2) that there is between Z_0 and Z_A a unique point Z_1 of Δ such that $\mathcal{H}(Z_0 , Z_1) = 0$.

We see that $\mathcal{H}$ ***is connected*** for $\tau \leqslant \tau_0$. According to (VIII.9.7) one has for $Z \in \mathcal{H}$ in the neighborhood of Z_0

$$(dS / d\tau)_{\mathcal{H}} < 0 \qquad\qquad (dS / dp)_{\mathcal{H}} > 0$$

It follows from Sect. (VIII.7) d that these inequalities are true over the curve of Hugoniot for all $\tau < \tau_0$.

Proposition. *Under the hypotheses concerning the equation of state and if* $\tau'_S > 0$, *Hugoniot's curve is connected. To* $m < m_0$ *corresponds a unique point* $Z_1 \neq Z_0$ *of* $\mathcal{H}$ *(with* $\alpha_1 \alpha_0 > 0$*) such that* $m = (\overline{q}_1 - \overline{q}_0)(\tau_1 - \tau_0)$. *One has over* $\mathcal{H}$ *the inequalities*

(VIII.10.3) $\quad (dS / d\tau)_{\mathcal{H}} < 0 \qquad (dS / dp)_{\mathcal{H}} > 0$

b) Consider an initial state Y_0 and , for $a = a(Y_0)$ the straight line Δ_a of P issued from Z_0 with slope m_a . We suppose

(VIII.10.4) $\quad m_a < m_0$

that is, according to (VIII.8.3) , $\alpha_0 \, P(\ell)(Z_0) > 0$. To each shoch $Y_0 \rightarrow Y_1$ corresponds a point Z_1 , the intersection of $\mathcal{H}$ and $\Delta_a \neq Z_0$.

Conversely, let Z_1 be the point (with VIII.10.4) $\neq Z_0$ where Δ_a meets $\mathcal{H}$; Z_1 being known, $\tau_1 < \tau_0$, $p_1 > p_0$ and ψ_1 are known ; f_1 is given by the Hugoniot relation ; $r_1 = f_1 / \tau_1$ being known, (VII.3.2) gives the value of $u_1^\alpha \ell_\alpha$ and (VII.3.3) the value of $h_1^\alpha \ell_\alpha$; $|h_1|^2$ may be given by the value of ψ_1 :

$$|h_1|^2 (\ell^\alpha \ell_\alpha - (u_1^\alpha \ell_\alpha)^2) + (h_1^\alpha \ell_\alpha)^2 = \psi_1^2 \ell^\alpha \ell_\alpha$$

It is easy to verify that these quantities satisfy the five invariance relations (VII.3.2).. (VII.3.6). We see that each point $Z_1 \neq Z_0$ belonging to $\mathcal{H}$ and Δ_a (with (VIII.10.4)) defines a unique non trivial solution of the shock equations such that $\alpha_0 \alpha_1 > 0$. We have the following theorem.

Theorem. *Suppose* $\tau'_S > 0$ *and the other hypotheses concerning the equation of state : Let* Y_0 *be a state satisfying* $\alpha_0\, P(\ell)_0 > 0$ *(that is* $v_0^{RM} < v_0^{\Sigma}$ *for a rapid shock ,* $v_0^{SM} < v_0^{\Sigma}$ *for a slow shock). Then there exists a corresponding unique non trivial solution of the shock equations such that* $v_1^{A} < v_1^{\Sigma}$ *for a rapid shock and* $v_1^{\Sigma} < v_1^{A}$ *for a slow shock. The solution satisfies the thermodynamic inequalities for the shocks* (VIII.6.3) (VIII.6.4) *and the inequalities of the theorem of Section* VIII.8.

We note that, for $H > 0$, there exist states Y_0 with $m_0 > 0$ that may give slow shocks, but not weak shocks ($m_a < 0$)

VIII.11. - The case $\tau'_S < 0$. [11].

a) Consider the isentropic curves defined by (VIII.7.2), let $\bar{q} = F(\tau,S)$. For a τ fixed, $\bar{q}$ is decreasing when S is increasing (according to (VIII.7.6) , $F'_S < 0$) and we have the relative location of the strictly concave isentropic curves. Let $Z_1 \neq Z_0$ be an arbitrary point of $\mathcal{H}$ ($\tau_1 < \tau_0$) and Δ the straight line (Z_0 , Z_1) with slope m . According to $\mathcal{H}(Z_0 , Z_0) = \mathcal{H}(Z_0 , Z_1) = 0$, $\mathcal{H}$ is stationary at one point Z_s of Δ between Z_0 and Z_1. The same is true for S and Z_s is a strict minimum for S and $\mathcal{H}(Z_0 , Z)$ along Δ .

Suppose that Z_1 is in the ***upper region*** determined by $\mathcal{S}_0$; Δ intersects $\mathcal{S}_0$ at one point Z_A such that Z_1 is between Z_0 and Z_A . When Z goes from Z_0 to Z_A , $\mathcal{H}(Z_0 , Z)$ is decreasing from Z to Z_s , has a minimum at Z_s and is increasing thereafter up to the value $\mathcal{H}(Z_0 , Z_A) < 0$ according to (VIII.9.1). We have then a contradiction with $\mathcal{H}(Z_0 , Z_1) = 0$. We see that ***each point*** $Z_1 \neq Z_0$ ***of*** $\mathcal{H}$ (with $\tau_1 < \tau_0$) ***is in the lower region determined by*** $\mathcal{S}_0$. Therefore the isentropic curve $\mathcal{S}_1$ of Z_1 corresponds to an entropy $S_1 > S_0$.

Now, when Z goes from Z_0 to Z_1 , S is decreasing from Z_0 to Z_s , has a minimum at Z_s and is increasing thereafter up to the value $S_1 > S_0$. Thus Δ intersects $\mathcal{S}_0$ at one point Z_A , between Z_0 and Z_1 . We know that necessarily, according to (VIII.7.10)

(VIII.11.1) $\qquad m < m_0$

Each point $\neq Z_0$ ***of*** $\mathcal{H}$ $(\tau < \tau_0)$ ***belongs to the open region defined by the isentropic curve*** $\mathcal{S}_0$ of Z_0 ***and by the tangent to*** $\mathcal{S}_0$ ***at*** Z_0 .

If $\mathcal{H}(Z_0 , Z)$ is expressed in terms of the variables $(\tau , \overline{q})$, study $\partial\mathcal{H} / \partial\overline{q}$. For a fixed τ , one has according to (VIII.2.1)

$$d\mathcal{H} = 2 f \Theta dS + (\tau - \tau_0) d\overline{q}$$

where according to (VIII.7.6)

$$d\overline{q} = - \tau'_S \tau'^{-1}_p \, dS \qquad\qquad dS = - \tau'^{-1}_S \tau'_p d\overline{q}$$

It follows

$$\partial\mathcal{H} / \partial\overline{q} = - 2 f \Theta \tau'^{-1}_S \tau'_p + (\tau - \tau_0) < 0$$

Therefore $\bar{q}$ is a regular function of τ, without singular points, along a connected component $\mathcal{H}_0$ of $\mathcal{H}$.

b) ***We consider only in the following the case*** $m_0 < 0$ (existence of weak shocks). Let $Z_1 \neq Z_0$ be a point of $\mathcal{H}_0$. With the notations of Section (VIII.7) one has

$$Q(m, Z_1) < 0$$

that is, according to (VIII.3.2)

$$(d\bar{q} / d\tau)_{\mathcal{S}_1} (Z_1) < m$$

But one has $m < m_0 < 0$. It follows

(VIII.11.2) $$(d\bar{q} / d\tau)_{\mathcal{S}_1} (Z_1) < 0$$

Study the sign of $(d\bar{q} / d\tau)_{\mathcal{H}_0} (Z_1)$ and suppose

$$(d\bar{q} / d\tau)_{\mathcal{H}_0} (Z_1) \geqslant 0$$

We deduce from (VIII.2.1)

(VIII.11.3) $$\{2 f \Theta (dS / d\tau)_{\mathcal{H}_0}\}(Z_1) = -(\tau_1 - \tau_0)(d\bar{q} / d\tau)_{\mathcal{H}_0} (Z_1) + (\bar{q}_1 - \bar{q}_0) > 0$$

since $(\bar{q}_1 - \bar{q}_0) = m(\tau_1 - \tau_0) > 0$. According to Section (VIII.7) d, S is a strictly monotone function of τ over $\mathcal{H}_0$ and we see that $(dS / d\tau)_{\mathcal{H}}$

is a strictly monotone function of τ over $\mathcal{H}_0$ and we see that $(dS/d\tau)_{\mathcal{H}}$ and $(d\overline{q}/d\tau)_{\mathcal{H}_0}(Z_1)$ are contradictory with the location of the isotropic curves.

It follows that $(d\overline{q}/d\tau)_{\mathcal{H}_0}$ *is always strictly negative.* According to the a , when τ is increasing, $\mathcal{H}_0$ ends necessarily at Z_0. Therefore, under the hypothesis $m_0 < 0$, ***the Hugoniot curve is connected*** and the same argument as in Section (VIII.10) shows that ***one has on*** $\mathcal{H}$ ***the inequalities***

(VIII.11.4) $\qquad (dS/d\tau)_{\mathcal{H}} < 0 \qquad\qquad (dS/dp)_{\mathcal{H}} > 0$

According to the theorem of Section (VIII.6), one has for a shock

$$S_1 < S_0 \qquad\qquad \tau_1 < \tau_0$$

and the inequality $(dS/dp)_{\mathcal{H}} > 0$ implies $p_1 > p_0$ and, according to Hugoniot's relation, $f_1 > f_0$.

The thermodynamic inequalities (VIII.6.4) ***are true again under the hypothesis*** $\tau_S' < 0$ ***if*** $m_0 < 0$.

c) Suppose always $\tau_S' < 0$ and $m_0 < 0$, and compare the values of $(d\overline{q}/d\tau)_{\mathcal{H}}$ and $(d\overline{q}/d\tau)_{\mathcal{S}}$. We deduce from (VIII.2.1) that

$$2f\,\Theta\,(dS/d\tau)_{\mathcal{H}} + (\tau - \tau_0)\,(d\overline{q}/d\tau)_{\mathcal{H}} - (\overline{q} - \overline{q}_0) = 0$$

But it follows from the equation of state and from the definition of $\overline{q}$ that

$$(dS/d\tau)_{\mathcal{H}} = \tau_S'^{-1} - \tau_p' \tau_S'^{-1} \{(d\bar{q}/d\tau)_{\mathcal{H}} + c^2 \mu \psi^2 \alpha^{-1}\}$$

One obtains

$$\{(\tau - \tau_0) - 2 f \Theta \tau_p' \tau_S'^{-1}\}(d\bar{q}/d\tau)_{\mathcal{H}} + 2 f \Theta \tau_p' \tau_S'^{-1} (\tau_p'^{-1} - c^2 \mu \psi^2 \alpha^{-1}) -$$

$$- (\bar{q} - \bar{q}_0) = 0$$

where

$$(d\bar{q}/d\tau)_{\mathcal{S}} = \tau_p'^{-1} - c^2 \mu \psi^2 \alpha^{-1}$$

We may write :

$$\{1 - (\tau - \tau_0) - 2 f \Theta)^{-1} \tau_p'^{-1} \tau_S'\}(d\bar{q}/d\tau)_{\mathcal{H}} =$$

$$= (d\bar{q}/d\tau)_{\mathcal{S}} - (\bar{q} - \bar{q}_0)(2 f \Theta)^{-1} \tau_p'^{-1} \tau_S'$$

let :

(VIII.11.5) $$(d\bar{q}/d\tau)_{\mathcal{H}} - (d\bar{q}/d\tau)_{\mathcal{S}} =$$

$$(2f \Theta)^{-1} \tau_p'^{-1} \tau_S' (\tau - \tau_0)\{(d\bar{q}/d\tau)_{\mathcal{H}} - (\bar{q} - \bar{q}_0)/(\tau - \tau_0)\}$$

Let $Z \neq Z_0$ be an arbitrary point of $\mathcal{H}$ (with $\tau < \tau_0$), Δ the straight line (Z_0, Z) with slope $m = (\bar{q} - \bar{q}_0)/(\tau - \tau_0)$, where $m < m_0 < 0$. Suppose

$$(d\bar{q}/d\tau)_{\mathcal{H}}(Z) \leqslant (d\bar{q}/d\tau)_{\mathcal{S}}(Z)$$

One has, according to Section (VIII.7) c

$$Q(m, Z) = (d\bar{q} / d\tau)_{\mathcal{S}} (Z) - m < 0$$

and it follows

$$(d\bar{q} / d\tau)_{\mathcal{H}} < m$$

It follows from (VIII.11.5) that the right hand side is positive and we have a contradiction. Therefore we have

(VIII.11.6) $$(d\bar{q} / d\tau)_{\mathcal{S}} (Z) < (d\bar{q}/d\tau)_{\mathcal{H}} (Z)$$

d) Let Z_1 and Z_2 (with $\tau_2 < \tau_1$) be two arbitrary points of $\mathcal{H}$, where Z_2 is in the neighborhood of Z_1. Consider now the chord $\Delta = (Z_1, Z_2)$ with slope m. One has $\mathcal{H}(Z_0, Z_1) = \mathcal{H}(Z_0, Z_2) = 0$. Therefore $\mathcal{H}(Z_0, Z)$ is stationary at one point Z_σ of the segment (Z_1, Z_2).

One has on the straight line Δ

$$\bar{q} - \bar{q}_1 = m (\tau - \tau_1) \qquad \bar{q} - \bar{q}_0 = \bar{q}_1 - \bar{q}_0 + m(\tau - \tau_1) \qquad d\bar{q} = m\, d\tau$$

We deduce from (VIII.2.1)

$$d\mathcal{H} = 2 f \Theta\, dS + (\tau - \tau_0) m\, d\tau - (\bar{q}_1 - \bar{q}_0 + m\tau - m\tau_1)\, d\tau$$

Denote by m_1 the slope of the straight line (Z_0, Z_1). One has after simplifications

(VIII.11.7) $$d\mathcal{H} = 2 f \Theta\, dS + (\tau_1 - \tau_0)(m - m_1)\, d\tau$$

If $m < m_1$, one has $(dS/d\tau)_\Delta (Z_\sigma) < 0$. If $(dS/d\tau)_\Delta \neq 0$ along the segment (Z_1, Z_2) , this derivative is negative and one has $S_2 > S_1$; m being < 0 , one obtains a contradiction with the relative location of the isentropic curves given in a . This hypotheses are impossible.

If $m > m_1$, one has $(dS/d\tau)_\Delta (Z_\sigma) > 0$. If $(dS/d\tau)_\Delta \neq 0$ along the segment (Z_1, Z_2) , this derivative is positive and one has in Z_1

$$(\tau'_S (dS/d\tau)_\Delta)(Z_1) < 0$$

and, according to (VIII.3.3) :

$$(Q(m))(Z_1) > 0 \qquad \text{(VIII.11.8)}$$

If $m = m_1$ S is stationary at Z_σ

We are led to study the case where $\mathcal{S}$ is stationary at a point Z_8 of the segment (Z_1, Z_2) that is a strict minimum, according to the lemma of Section (VIII.7) . If such is the case, one has again the inequality (VIII.11.8)

In all the cases, one has

$$(Q(m))(Z_1) > 0$$

that is

$$(d\bar{q}/d\tau)_{\mathcal{S}} (Z_1) > m$$

It follows from (VIII.11.6).

It follows from (VIII.11.6).

(VIII.11.9) $\qquad (d\bar{q}/d\tau)_{\mathcal{H}}\,(Z_1) > m$

where m is the slope of the segment $(Z_1\, Z_2)$. We deduce from (VIII.11.9) that $\mathcal{H}$ ***is concave.***

If Z_2 is an arbitrary point of $\mathcal{H}$ and m_2 the slope of the straight line $(Z_0\,,Z_2)$, each straight line Δ issued from Z_0 of slope m such that

$$m_2 < m < m_0$$

intersects $\mathcal{H}$ in one and only one point Z.

We have proved that $\mathcal{H}$ is connected and concave and that we have on $\mathcal{H}$ the inequalities

$$(dS/d\tau)_{\mathcal{H}} < 0 \qquad\qquad (dS/dp)_{\mathcal{H}} > 0$$

The same argument as in Section (VIII.10) gives the following theorem [11].

Theorem. *Suppose* $\tau_S' < 0$, $m_0 < 0$ *(existence of weak shocks) and the other hypotheses concerning the equation of state. Let* Y_0 *be a state satisfying* $\alpha_0\, P(\ell)_0 > 0$ *(that is* $v_0^{RM} < v_0^{\Sigma}$ *for a rapid shock,* $v_0^{SM} < v_0^{\Sigma}$ *for a slow shock). If* Z_2 *is a point of* $\mathcal{H}$ *and if* $(q_2 - q_0)(\tau_2 - \tau_0)^{-1} \leqslant c^2 a^2 / \ell^\alpha \ell_\alpha$, *there exists a corresponding unique non trivial solution of the shock equations such that* $v_1^A < v_2^{\Sigma}$ *for a rapid shock and* $v_1^{\Sigma} < v_1^A$, *for a slow shock. The solution satisfies the thermodynamic* *inequalities of the theorem of Section* VIII.8.

The case $m_0 \geqslant 0$ remains open.

VIII.12. - Conclusions.

Under the general hypotheses $\tau'_p < 0$, $\tau''_{p^2} > 0$, $\tau'_S \neq 0$ we have obtained the following results :

- orientation of the shock wave fronts in the time
- thermodynamic inequalities for the shocks (in particular each shock wave is a compression wave
- relation locations of the speeds of the shock fronts with respect to the speeds of the magnetosonic and Alfven waves.

We have proved also by two different methods and under lightly different hypotheses, for $\tau'_S > 0$ and $\tau'_S < 0$, existence and uniqueness theorems for non trivial solutions of the shock equations.

ANNEX I. - SHOCK WAVES AND ALFVEN WAVES

AI.1. - Singular shocks. [9]

Consider a non tangential shock wave that is not an Alfven shock and let Σ be the wave front of this shock. We know that $\ell^\alpha \ell_\alpha$ is < 0 . The two following vectors are invariant under the shock.

(AI.1.1) $$V^\beta = (h^\alpha \ell_\alpha)\ u^\beta - (a/r)\ h^\beta$$

(AI.1.2) $$W^\beta/\mu = \beta^2(a/r)\ u^\beta - (q/\mu)\ \ell^\beta - (h^\alpha \ell_\alpha)\ h^\beta \qquad ,$$

with $$\beta^2 = (c^2 rf + \mu\,|\,h\,|^2)$$

a) One has the following lemma

Lemma. *There exists always one direction* n *at least that is orthogonal to* u_0 , h_0 , u_1 , h_1 .

In fact, consider the different cases.

1) *If* $\alpha_0\,\alpha_1 \neq 0$, there exists one direction n at least that is orthogonal to the vectors ℓ , V and X with (see (VII.2.4))

(AI.1.3) $$X^\beta = a\, r_0\, \alpha_0\, v_0^\beta + \mu\,(r_0\,/\,a)\ h_0^\alpha \ell_\alpha\, V^\beta =$$

$$= a\, r_1\, \alpha_1\, v_1^\beta + \mu\,(r_1\,/a)(h_1^\alpha \ell_\alpha)\, V^\beta$$

From (AI.1.3), it follows that $\mathbf{n}$ is orthogonal also to v_0 and v_1, and thus to $\mathbf{u}_0$ and $\mathbf{u}_1$. But, according to (AI.1.1), $\mathbf{n}$ is orthogonal also to $\mathbf{h}_0$ and $\mathbf{h}_1$.

2) *If* $\alpha_0 \neq 0$, $\alpha_1 = 0$, one has $\chi_0 = 0$ and it follows from the lemma 2 of Section (VII.2) that ℓ belongs to the 2-plane π_0 defined by $(\mathbf{u}_0, \mathbf{h}_0)$. The vector V is orthogonal to ℓ in the 2-plane π_0. There exists one direction $\mathbf{n}$ at least orthogonal to ℓ, V and $\mathbf{u}_1$. This direction is orthogonal to π_0; $\mathbf{n}$ being orthogonal to V and $\mathbf{u}_1$ is orthogonal to $\mathbf{h}_1$.

3) *If* $\alpha_0 = 0$, $\alpha_1 \neq 0$, it is sufficient to exchange the roles of the induces 0 and 1. Our lemma is proved.

b) Introduce at $x \in \Sigma$ an infinitesimal perturbation of the state Y_0 before the shock. One obtains an infinitesimal perturbation of the state Y_1 after the shock. If one denote here by δ the perturbation operator, the two perturbations are connected by the relations given by the differentiation of the fundamental equations of shock

$$\text{(AI.1.4)} \quad \begin{cases} [(\delta r)\, u^\alpha + r\, \delta u^\alpha\,]\, \ell_\alpha = 0 \\[2ex] [\delta(h^\alpha \ell_\alpha)\, u^\beta + (h^\alpha \ell_\alpha)\, \delta u^\beta - \delta(a/r)\, h^\beta - (a/r)\, \delta h^\beta\,] = 0 \\[2ex] [\delta(\beta^2 (a/r))\, u^\beta + \beta^2 (a/r)\, \delta u^\beta - (\delta q/\mu)\, \ell^\beta - \delta(h^\alpha \ell_\alpha)\, h^\beta \\[1ex] \qquad - (h^\alpha \ell_\alpha)\, \delta h^\beta\,] = 0 \end{cases}$$

Let $\{ \mathbf{e}_{(\alpha)} \}$ be an orthonormalized frame in x, such that $\mathbf{e}_{(2)}$ is colinear to $\boldsymbol{\ell}$ and $\mathbf{e}_{(3)}$ to $\mathbf{n}$. In the frame, one has

$$u_0^3 = 0 \qquad h_0^3 = 0 \qquad u_1^3 = 0 \qquad h_1^3 = 0$$

Suppose that only δu_0^3 and δh_0^3 are $\neq 0$ before the shock. One has $\delta(u_0^\alpha \ell_\alpha) = 0$, $\delta(h_0^\alpha \ell_\alpha) = 0$, $\delta p_0 = 0$, $\delta S_0 = 0$ and so $\delta \tau_0 = 0$, $\delta f_0 = 0$, $\delta r_0 = 0$, $\delta |h_0|^2 = 0$.

We suppose that $\delta p_1 = 0$, $\delta S_1 = 0$ and so $\delta \tau_1 = 0$, $\delta f_1 = 0$. By differentiation of (VII.3.2) (VII.3.3) one obtains $\delta(u_1^\alpha \ell_\alpha) = 0$, $\delta h_1^\alpha \ell_\alpha = 0$. From $\delta \tau_1 = 0$, it follows $\delta \alpha_1 = 0$ and, by differentiation of (VII.3.5), one has $\delta \chi_1 = 0$ and so $\delta |h_1|^2 = 0$. Under our hypotheses, it is easy to see that all the shock relations (AI.1.4) are satisfied if we add the relations corresponding to $\beta = 3$, that is

$$\text{(AI.1.5)} \quad (h_1^\alpha \ell_\alpha)\, \delta u_1^3 - (a/r_1)\, \delta h_1^3 = (h_0^\alpha \ell_\alpha)\, \delta u_0^3 - (a/r_0)\, \delta h_0^3$$

$$\text{(AI.1.6)} \quad \beta_1^2 (a/r_1)\, \delta u_1^3 - (h_1^\alpha \ell_\alpha)\, \delta h_1^3 = \beta_0^2 (a/r_0)\, \delta u_0^3 - (h_0^\alpha \ell_\alpha)\, \delta h_0^3$$

The thermodynamic variables being without perturbations, it follows that, for the state before the shock (resp. after the shock) such perturbations correspond to infinitesimal Alfven shocks, that are to Alfven waves

We have seen that these Alfven waves are of two types A and B with

$$A^\alpha = \beta u^\alpha + h^\alpha \qquad\qquad B^\alpha = \beta u^\alpha - h^\alpha$$

Consider, in the state Y_0 before the shock, an Alfven wave of type A. The vector A_0^α being invariant at the traverse of this infinitesimal wave, such a wave carries in x a perturbation $(\delta u_{0A}^3, \delta h_{0A}^3)$ such that

(AI.1.7) $$\beta_0\, \delta u_{0A}^3 + \delta h_{0A}^3 = 0$$

In the same way, an Alfven wave of type B carries in x a perturbation $(\delta u_{0B}^3, \delta h_{0B}^3)$ such that

(AI.1.8) $$\beta_0\, \delta u_{0B}^3 - \delta h_{0B}^3 = 0$$

It is easy to see that the superposition in x of a wave of type A and of a wave of type B gives an arbitrary perturbation $(\delta u_0^3, \delta h_0^3)$ with

$$\delta u_0^3 = \delta u_{0A}^3 + \delta u_{0B}^3 \qquad \delta h_0^3 = \delta h_{0A}^3 + \delta h_{0B}^3$$

One has

$$A_0^\alpha = \beta\, u_\alpha^0 + h_0^\alpha \qquad B_0^\alpha = \beta_0\, u_0^\alpha - h_0^\alpha$$

It follows :

(AI.1.9) $$A_0^\alpha \ell_\alpha = \beta_0\,(a/r) + h_0^\alpha \ell_\alpha \qquad B_0^\alpha \ell_\alpha = \beta_0\,(a/r) - h_0^\alpha \ell_\alpha$$

One obtains by multiplication

$$(A_0^\alpha \ell_\alpha)(B_0^\alpha \ell_\alpha) = \beta_0^2\,(a^2 / r_0^2) - (h_0^\alpha \ell_\alpha)^2 = a^2/\mu\ \alpha_0$$

In the following, ***we suppose ℓ oriented with respect to Σ from the state before the shock to the state after the shock.*** One has then $a < 0$. Suppose, for example, $b > 0$ and thus $(h_0^\alpha \ell_\alpha) > 0$ (resp $(h_1^\alpha \ell_\alpha) > 0$). According to (AI.1.9), the vector B_0 (resp. B_1) is oriented in the same way as ℓ with respect to Σ . For the vector A_0 (resp. A_1) its orientation is the orientation of ℓ or the opposite orientation according to $\alpha_0 > 0$ or $\alpha_0 < 0$ (resp. $\alpha_1 > 0$ or $\alpha_1 < 0$). For $\alpha = 0$, A is tangent to Σ . If b is < 0 , the roles of the vectors A and B are exchanged.

c) Analyse now the case of the ***singular shocks.*** Suppose first

$$\alpha_1 = 0 \qquad\qquad \alpha_0 > 0$$

In the state Y_0 , the Alfven waves of the two types may generate at $x \in \Sigma$ an arbitrary perturbation $(\delta u_0^3 , \delta h_0^3)$ and in the relations (AI.1.5) , (AI.1.6) one has since $\alpha_1 = 0$

(AI.1.10) $$\beta_1^2 (a^2 / r_1^2) - (h_1^\alpha \ell_\alpha)^2 = 0$$

It follows from (AI.1.5) , (AI.1.6) and from (AI.1.10)

$$\beta_1^2 (a/r_1) ((h_0^\alpha \ell_\alpha) \delta u_0^3 - (a/r_0) \delta h_0^3)$$

$$- (h_1^\alpha \ell_\alpha) (\beta_0^2 (a/r_0) \delta u_0^3 - (h_0^\alpha \ell_\alpha) \delta h_0^3) = 0$$

The relations (AI.1.5), (AI.1.6) have a solution for all perturbation $(\delta u_0^3 , \delta h_0^3)$ if and only if one has

(AI.1.11) $$\beta_1^2 (a/r_1)(h_0^\alpha \ell_\alpha) - \beta_0^2 (a/r_0)(h_1^\alpha \ell_\alpha) = 0$$

and

(AI.1.12) $$\beta_1^2 (a/r_1)(a/r_0) - (h_1^\alpha \ell_\alpha)(h_0^\alpha \ell_\alpha) = 0$$

The relation (AI.1.11) may be written

$$\frac{\beta_1^2 (a/r_1)}{\beta_0^2 (a/r_0)} = \frac{h_1^\alpha \ell_\alpha}{h_0^\alpha \ell_\alpha} = k$$

and (AI.I.12) gives then $k\alpha_0 = 0$, let $k = 0$; that is impossible. ***The shock wave is not compatible with the Alfven waves***

d) Suppose now

$$\alpha_1 < 0 \qquad \alpha_0 = 0$$

One has then $\mu H = c^2 \tau_0$ and H is > 0 . According to $\alpha_0 = 0$, one has

(A.I.I.13) $$\beta_0 (a/r_0) + h_0^\alpha \ell_\alpha = 0$$

Before the shock, a wave of type B and a wave of type A ***tangent to*** Σ may create at $x \in \Sigma$ an arbitrary perturbation $(\delta u_0^3, \delta h_0^3)$. But only the wave of type A for the state Y_0 carrying a perturbation $(\delta u_0^3, \delta h_0^3)$ satisfying:

(AI.1.14) $$\beta_0 \, \delta u_{0A}^3 + \delta h_{0A}^3 = 0$$

and an Alfven wave of type B for the state Y_1 carrying a perturbation $(\delta u^3_{1B}, \delta h^3_{1B})$ with

(AI.1.15) $$\beta_1 \, \delta u^3_{1B} - \delta h^3_{1B} = 0$$

can go away.

The relations (AI.1.5), (AI.1.6) connect this last perturbation to a perturbation $(\hat{\delta} u^3_0, \hat{\delta} h^3_0)$ before the shock with

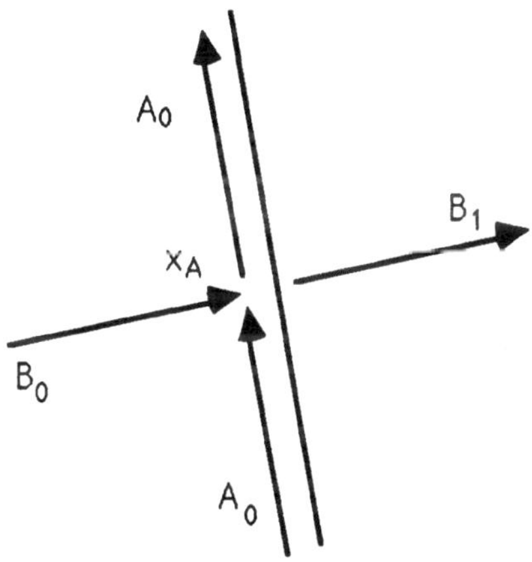

$$((h^\alpha_1 \ell_\alpha) - \beta_1 (a/r_1)) \, \delta u^3_{1B} = (h^\alpha_0 \ell_\alpha) \, \hat{\delta} \, u^3_0 - (a/r_0) \, \hat{\delta} \, h^3_0$$

$$\beta_1 (\beta_1 (a/r_1) - (h^\alpha_1 \ell_\alpha) \, \delta u^3_{1B} = \beta^2_0 (a/r_0) \, \hat{\delta} \, u^3_0 - (h^\alpha_0 \ell_\alpha) \, \hat{\delta} \, h^3_0$$

A perturbation $(\hat{\delta} u^3_0, \hat{\delta} h^3_0)$ can be transmitted at the traverse of Σ and so can go away if and only if

(AI.1.16) $(\beta_1(h_0^\alpha \ell_\alpha) + \beta_0^2 (a/r_0))\, \hat{\delta}\, u_0^3 - (\beta_1 (a/r_0) + (h_0^\alpha \ell_\alpha))\, \hat{\delta}\, h_0^3 = 0$

We shall prove in a moment that

(AI.1.17) $$\beta_1 (a/r_0) + h_0^\alpha \ell_\alpha \neq 0$$

The shock wave is compatible with an arbitrary perturbation $(\delta u_0^3, \delta h_0^3)$ if there is always a decomposition

$$\delta u_0^3 = \hat{\delta} u_0^3 + \delta u_{0A}^3$$

$$\delta u_0^3 = \hat{\delta} h_0^3 + \delta h_{0A}^3$$

with

$$\hat{\delta}\, h_0^3 = \nu\, \hat{\delta}\, u_0^3 \qquad \delta h_{0A}^3 = -\beta_0\, \delta u_{0A}^3$$

where, according to (AI.1.16)

(AI.1.18) $$\nu = \frac{\beta_1 (h_0^\alpha \ell_\alpha) + \beta_0^2 (a/r_0)}{\beta_1 (a/r_0) + (h_0^\alpha \ell_\alpha)}$$

We are led to study the linear system in $\hat{\delta} u_0^3$, δu_{0A}^3 :

(AI.1.19) $$\begin{aligned} \hat{\delta} u_0^3 + \delta u_{0A}^3 &= -\delta u_0^3 \\ \nu\, \hat{\delta}\, u_0^3 - \beta_0\, \delta u_0^3 &= \delta h_0^3 \end{aligned}$$

One has for the determinant of (AI.1.19)

$$\nu + \beta_0 = \frac{\beta_1 (h_0^\alpha \ell_\alpha) + \beta_0^2 (a/r_0) + \beta_0 (\beta_1 (a/r_0) + h_0^\alpha \ell_\alpha)}{\beta_1 (a/r_0) + (h_0^\alpha \ell_\alpha)}$$

where $(h_0^\alpha \ell_\alpha)$ is given by (AI.1.13). One obtains $(\nu + \beta_0) = 0$ and (AI.1.18) admits no solution for every values of the right-hand sides of these equations. Therefore ***the shock wave is not compatible with the Alfven waves.***

We will prove now that $\beta_1 (a/r_0) + (h_0^\alpha \ell_\alpha) \neq 0$. If not is the case, one has $\beta_1 = \beta_0$ or

$$c^2 r_1 f_1 + \mu |h_1|^2 = c^2 r_0 f_0 + \mu |h_0|^2$$

This formula may be written

$$r_1^2 (c^2 \tau_1 + \mu (|h_1|^2 / r_1^2)) = r_0^2 (c^2 \tau_0 + \mu |h_0|^2 / r_0^2)$$

or

$$r_1^2 (\alpha_1 + \mu ((h_1^\alpha \ell_\alpha)^2 / a^2)) = r_0^2 \mu ((h_0^\alpha \ell_\alpha)^2 / a^2) =$$

$$= r_0^2 \mu ((h_1^\alpha \ell_\alpha)^2 / a^2)(f_1^2 / f_0^2)$$

It follows

$$(1/\tau_1^2) \alpha_1 = \mu ((h_1^\alpha \ell_\alpha)^2 / a^2) ((1/\tau_0^2) - (1/\tau_1^2))$$

let since $\alpha_1 = c^2 (\tau_1 - \tau_0)$

$$(h_1^\alpha \ell_\alpha)^2 / a^2) (\tau_0 + \tau_1) = H\tau_0 = \{(h_1^\alpha \ell_\alpha)^2 / a^2) - (|h_1|^2 / r_1^2)\} \tau_0$$

One obtains after simplifications

$$\{ (h_1^\alpha \ell_\alpha)^2 / a^2 \} \tau_1 + \{ |h_1|^2 / r_1^2 \} \tau_0 = 0$$

It follows $h_1 = 0$ that is $H = 0$. But H is > 0

One has

Theorem. - *A non tangential shock wave that is not an Alfven shock, but such that* $\alpha_0 \alpha_1 = 0$ *is not compatible with the Affven waves.*

The singular shocks (Section (VII.3)) *are physically forbidden* as unstable with respect to the infinitesimal Alfven waves.

AI.2. - Compatibility between schock waves and Alfven waves [9].

a) Consider a non tangential shock wave that is not an Alfven shock and suppose first that one has

$$0 < \alpha_1 < \alpha_0$$

The Alfven waves of types A and B in the state Y_0 are creating in $x \in \Sigma$ an arbitrary perturbation $(\delta u_0^3, \delta h_0^3)$; α being > 0, the system (AI.1.5) (AI.1.6) admits a unique solution defining a perturbation of the state Y_1 that can go away according to the Alfven waves A_1 and B_1 corresponding to this state.

One has compatibility of the shock wave with the Alfven waves.

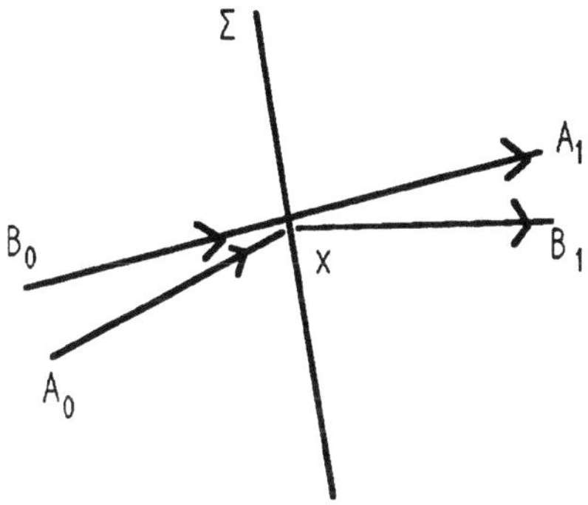

$0 < \alpha_1 < \alpha_0$

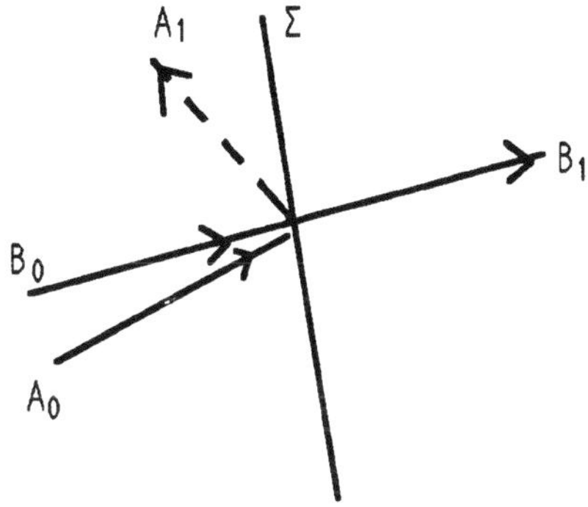

$\alpha_1 < 0 < \alpha_0$

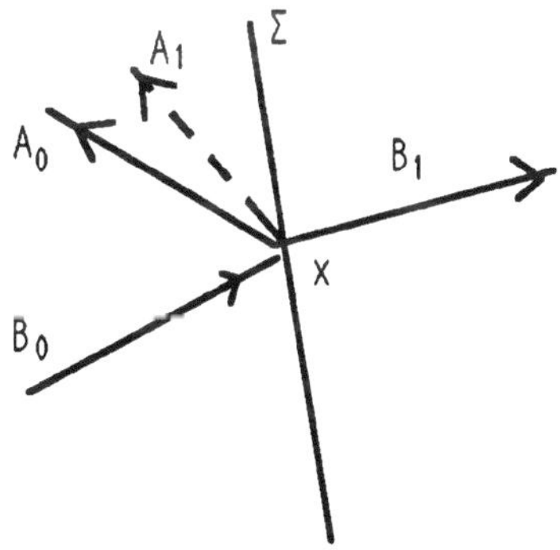

$\alpha_1 < \alpha_0 < 0$

b) Suppose now

$$\alpha_1 < 0 < \alpha_0$$

The Alfven waves of types A and B in the state Y_0 are always creating in $x \in \Sigma$ an arbitrary perturbation $(\delta u^3_0 , \delta h^3_0)$. But only an Alfven waves of type B carrying a perturbation satisfying

$$\beta_1 \, \delta u^3_{1B} - \delta h^3_{1B} = 0$$

can go away in the state Y_1 :

From (AI.1.5), (AI.1.6) , it follows

$$(h^\alpha_0 \ell_\alpha) \, \delta u^3_0 - (a/r_0) \, \delta h^3_0 = \{ ((\beta_1 (a/r_1) - (h^\alpha_1 \ell_\alpha)\} \, \delta u^3_{1B}$$

$$\beta^2_0 (a/r_0) \, \delta u^3_0 - (h^\alpha_0 \ell_\alpha) \, \delta h^3_0 = \beta_1 \{ \beta_1 (a/r_1) - (h^\alpha_1 \ell_\alpha)\} \, \delta u^3_{1\beta}$$

It is necessary that $(\delta u^3_0 , \delta h^3_0)$ satisfies the relation

(AI.2.1) $\{\beta^2_0 (a/r_0) + \beta_1 (h^\alpha_0 \ell_\alpha)\}\delta u^3_0 - \{ \beta_1 (a/r_0) + (h^\alpha_1 \ell_\alpha)\} \, \delta h^3_0 = 0$

If (AI.2.1) is an identity on $(\delta u^3_0 , \delta h^3_0)$, one has

$$\beta_1 = \beta_0 \qquad\qquad \beta_0 (a/r_0) + h^\alpha_0 \ell_\alpha$$

and thus $\alpha_0 = 0$. Therefore (AI.2.1) is not an identity

The shock wave is not compatible with the Alfven waves

c) Suppose finally

$$\alpha_1 < \alpha_0 < 0$$

Before the shock, one has only an Alfven wave of type B creating in $x \in \Sigma$ a perturbation $(\delta u^3_{0B}, \delta h^3_{0B})$ satisfying

(AI.2.2) $$\beta_0 \, \delta u^3_{0B} - \delta h^3_{0B} = 0$$

with

(AI.2.3) $$\beta_0 \, a/r_0 - (h^\alpha_0 \ell_\alpha) = 0$$

But only an Alfven wave of type A for the state Y_0 and an Alfven wave of type B for the state Y_1 can go away from x. The Alfven wave of type B for Y_1 corresponds before the shock to a perturbation $(\hat{\delta} u^3_0, \hat{\delta} h^3_0)$. We set

$$\delta u^3_{0B} = \hat{\delta} u^3_0 + \delta u^3_{0A} \qquad \delta h^3_{0B} = \hat{\delta} h^3_{0B} + \delta h^3_{0A}$$

According to (AI.2.3) one has

(AI.2.4) $$\beta_1 (a/r_0) + h^\alpha_0 \ell_\alpha \neq 0$$

It follows from (AI.1.18) that one has

$$\hat{\delta} h^3_0 = \nu \, \hat{\delta} u^3_0 \qquad \delta h^3_{0A} = -\beta_0 \, \delta u^3_{0A}$$

with

$$\nu = \frac{\beta_1 (h^\alpha_0 \ell_\alpha) + \beta_0^2 \, a/r_0)}{\beta_1 (a/r_0) + h^\alpha_0 \ell_\alpha}$$

We are led, as in Section (AI.1) to solve the linear system

$$\hat{\delta}\, u_0^3 + \delta\, u_{0A}^3 = \delta\, u_{0B}^3$$

$$\nu\, \delta u_0^3 - \beta_0\, \delta\, u_{0A}^3 = \beta_0\, \delta\, u_{0B}^3$$

with the determinant

$$\nu + \beta_0 = \frac{\beta_0 + \beta_1}{\beta_1\,(a/r_0) + h_0^\alpha \ell_\alpha}\; 2\,\beta_0\,(a/r_0)$$

is $\neq 0$

One has compatibility of the shock wave with the Alfven waves.

We have proved in this Annex I.

Theorem. - *A non tangential shock that is not an Alfven shock is compatible with the Alfven waves if and only if one has* $\alpha_0 \alpha_1 > 0$.

d) We see that the considered shocks may be decomposed in ***slow shocks*** such that

$$\alpha_1 < \alpha_0 < 0$$

rapid shocks such that

$$0 < \alpha_1 < \alpha_1$$

In the two cases, we have the main relation (VII.3.5) let

$$\chi_1 \alpha_1^2 = \chi_0 \alpha_0^2$$

For a slow shock $\alpha_1^2 > \alpha_0^2$ and $\chi_1 < \chi_0$. It follows $|h_1|^2 < |h_0|^2$. For a rapid shock, these inequalities are inverse. One has

Proposition. - *For a slow shock, the norm of the magnetic field is decreasing. For a rapid shock, this norm is increasing.*

ANNEX II - MAGNETOSONIC RAYS

A.II.1. - Directions of the rays.

a) From the fundamental equations of magnetohydrodynamics, it follows that for an infinitesimal wave with a front Σ non generated by the lines of flow $(u^\alpha \ell_\alpha \neq 0)$, one has for the operator δ of infinitesimal discontinuity

$$\delta(r u^\alpha \ell_\alpha) = 0 \qquad \delta\{(h^\alpha \ell_\alpha)\, u^\beta - (u^\alpha \ell_\alpha)\, h^\beta\} = 0 \qquad \delta(T^{\alpha\beta} \ell_\alpha) = 0$$

One obtains thus the invariance by δ of the scalar and the two vectors

$$\text{(AII.1.1)} \quad \begin{cases} a = r\, u^\alpha \ell_\alpha \neq 0 \\ V^\beta = (h^\alpha \ell_\alpha)\, v^\beta - (a/r)\, k^\beta \\ X^\beta = a\, r\, \alpha\, v^\beta + \mu\, (r/a)\, (h^\alpha \ell_\alpha)\, V^\beta \end{cases}$$

with the notations of (VII.1), (VII.2). We know that it follows that

$$b = f(h^\alpha \ell_\alpha) \qquad\qquad H = (h^\alpha \ell_\alpha)^2 / a^2 - |h|^2 / r^2$$

are also invariant.

Suppose that the wave is not an Afven wave $(\alpha \neq 0)$, but is a magnetosonic wave so that

(AII.1.2) $$P(\ell) = c^2 rf(\gamma-1)(u^\alpha \ell_\alpha)^4 + (c^2 rf + \mu |h|^2 \gamma)(u^\alpha \ell_\alpha)^2 \ell^\rho \ell_\rho -$$

$$- \mu (h^\alpha \ell_\alpha)^2 \ell^\rho \ell_\rho = 0$$

This equation may be written

(AII.1.3) $$((\gamma-1)/r^2)\, c^2 a^2 \tau + (c^2 \tau + \mu(|h|^2 / r^2)\gamma)\, \ell^\rho \ell_\rho -$$

$$- \mu(|h^\alpha \ell_\alpha|^2 / a^2)\, \ell^\rho \ell_\rho = 0$$

One has

$$((\gamma-1)/r^2)\, c^2 a^2 \tau + \mu(|h|^2 / r^2)(\gamma-1)\, \ell^\rho \ell_\rho + \mu\{|h|^\alpha / r^2) -$$

$$- ((h^\alpha \ell_\alpha)^2 / a^2\}\, \ell^\rho \ell_\rho + c^2 \tau\, \ell^\rho \ell_\rho = 0$$

that is

(AII.1.4) $$((\gamma-1)/r^2)(c^2 a^2 \tau + \mu |h|^2 \ell^\rho \ell_\rho) + \alpha\, \ell^\rho \ell_\rho = 0$$

where

$$c^2 a^2 \tau + \mu |h|^2\, \ell^\rho \ell_\rho \neq 0$$

b) According to Section (VI.9), one has

$$N^\beta = 2c^2 rf(\gamma-1)(u^\alpha \ell_\alpha)^3 v^\beta + (c^2 rf + \mu |h|^2 \gamma)(u^\alpha \ell_\alpha)\, \ell^\rho \ell_\rho v^\beta -$$

$$- \mu (h^\alpha \ell_\alpha) \ell^\rho \ell_\rho k^\beta$$

or

$$N^{\theta} = \{2((\gamma-1)/r^2)\, c^2 a^2 \tau + (c^2\tau + \mu(|h|^2/r^2)\,\gamma)\,\ell^{\rho}\ell_{\rho}\}\, a r v^{\theta} -$$

$$- \mu(h^{\alpha}\ell_{\alpha})\,\ell^{\rho}\ell_{\rho}\, k^{\theta} .$$

But according to (A.II.1.3), one has

$$N^{\theta} = \{((\gamma-1)/r^2)\, c^2 a^2 \tau + \mu((h^{\alpha}\ell_{\alpha})^2/a^2)\,\ell^{\rho}\ell_{\rho}\}\, a r v^{\theta} -$$

$$- \mu(h^{\alpha}\ell_{\alpha})\,\ell^{\rho}\ell_{\rho}\, k^{\theta}$$

The expression of $\mathbf{V}^{\theta}$ gives :

$$k^{\theta} = (r/a)(h^{\alpha}\ell_{\alpha})\, v^{\theta} - (r/a)\, \mathbf{V}^{\theta}$$

It follows

$$N^{\theta} = ((\gamma-1)/r^2)\, c^2 a^2 \tau\ \ a r v^{\theta} + \mu\, (r/a)(h^{\alpha}\ell_{\alpha})\,\ell^{\rho}\ell_{\rho}\,\mathbf{V}^{\theta}$$

We see from (AII.1.4) that

$$(\gamma-1)/r^2 = -\,\alpha\,\ell^{\rho}\ell_{\rho}\,/\,(c^2 a^2 \tau + \mu|h|^2\ell^{\rho}\ell_{\rho})$$

and we have :

$$N^{\theta} = -c^2 a^2 \tau\,\ell^{\rho}\ell_{\rho}\, a\alpha r v^{\theta}/(c^2 a^2 \tau + \mu|h|^2\ell^{\rho}\ell_{\rho}) +$$

$$+ \mu\,(r/a)(h^{\alpha}\ell_{\alpha})\,\ell^{\rho}\ell_{\rho}\,\mathbf{V}^{\theta}$$

We deduce from (AII.1.1)

$$N^{\beta} = -\{ c^2a^2\tau\, \ell^{\rho}\ell_{\rho} /(c^2 a^2 \tau + \mu | h |^2 \ell^{\rho}\ell_{\rho})\} (X^{\beta} - \mu (r/a)(h^{\alpha}\ell_{\alpha}) V^{\beta}) +$$

$$+ \mu(r/a)(h^{\alpha}\ell_{\alpha}) \ell^{\rho}\ell_{\rho} V^{\beta}$$

The direction of N^{β} is the direction of the vector

$$- c^2 a^2 \tau X^{\beta} + \mu(r/a)(h^{\alpha}\ell_{\alpha}) (2 c^2 a^2 \tau + \mu | h |^2 \ell^{\rho}\ell_{\rho}) V^{\beta}$$

Introduce $b = f h^{\alpha}\ell_{\alpha}$. One obtains, up to the factor τ , the vector

(AII.1.5) $$M^{\beta} = - c^2 a^2 X^{\beta} + 2 \mu (b/a) Q V^{\beta}$$

where

(AII.1.6) $$Q = (c^2 a^2 / \tau) + (1/2) (\mu | h |^2 \ell^{\rho}\ell_{\rho} / \tau^2)$$

AII.2. - Action of δ on the direction of the ray.

a) One has

$$\delta M^{\beta} = 2\mu (b/a) \delta Q V^{\beta}$$

In particular if $b = 0$, that is if the magnetic field is tangential, one has $\delta M^{\beta} = 0$ and the direction of N^{β} is invariant under δ .

Evaluate δQ by differentiation of

$$\tau^2 Q = c^2 a^2 \tau + (1/2) \mu | h |^2 \ell^{\rho}\ell_{\rho}$$

One has

$$\tau^2\,\delta Q + 2\tau\,\delta\tau\,(c^2 a^2/\tau + ((1/2)\,\mu\,|h|^2\,\ell^\rho\ell_\rho/\tau^2) = c^2 a^2\,\delta\tau + $$

$$+\frac{1}{2}\,\mu\,\delta|h|^2\cdot\ell^\rho\ell_\rho$$

that is

(AII.2.1) $\tau^2\,\delta Q = -\,c^2 a^2\,\delta\tau + (1/2)\,\ell^\rho\ell_\rho\,\mu\,\delta|h|^2 - \ell^\rho\ell_\rho\,\mu\,|h|^2\,\delta\tau/\tau$

According to (VII.3.6), one has

$$\delta(p + (1/2)\,\mu\,|h|^2) - (c^2 a^2/\ell^\rho\ell_\rho)\,\delta\tau = 0$$

or

$$-\,c^2 a^2\,\delta\tau + (1/2)\,\ell^\rho\ell_\rho\,\mu\,\delta|h|^2 = -\,\ell^\rho\ell_\rho\,\delta p$$

δS being null, (AII.2.1) may be written

(AII.2.2) $\tau^2\,\delta Q = -(\ell^\rho\ell_\rho/\tau)\,(\tau + \mu\,|h|^2\,\tau'_p)\,\delta p$

b) One has

$$c^2\,\tau'_p = -\,(\gamma - 1)/r^2$$

It follows

(AII.2.3) $\tau^2\,\delta Q = -(\ell^\rho\ell_\rho/c^2 r f)\,\{c^2 r f - (\gamma - 1)\,\mu\,|h|^2\}\,\delta p$

δQ is null if and only if

$$\gamma = (c^2 r f + \mu\,|h|^2)/\mu\,|h|^2$$

that is if

$$\text{(AII.2.4)} \qquad v^2 / c^2 = \mu\,|\,h\,|^2 / (c^2 rf + \mu\,|\,h\,|^2)$$

We have proved that for the rays associated to the sonic waves (Section (V.5)) and, in magnetohydrodynamics, to the Alfven waves, the direction is invariant under the operator δ of infinitesimal discontinuity, but that ***the direction of the rays associated to the magnetosonic waves are not in general invariant.***

We have only invariance if the magnetic field h is tangential or if the relation (AII.2.4) is satisfied, that is if ***we are in the special case*** (see Section (VI.8) and (VI.8.1)) ***for the cone*** $P(\ell) = 0$.

ANNEX III - CLASSICAL APPROXIMATION OF THE RELATIVISTIC SHOCK EQUATIONS

AIII.1. - The frame connected with the shock.

Consider a Minkowski space-time and in this space-time a non tangential shock of front Σ. In $x \in \Sigma$ introduce an orthonormalized frame $\{ e_{(\alpha)} \}$ such that $e_{(1)} = \ell$. We will study the Newtonian approximation of the relativistic main shock equations (VII.3.2) ... (VII.3.6). We have $a \neq 0$ and we put

(AIII.1.1) $\quad u^\alpha \ell_\alpha = w/c \qquad j = rw \qquad \ell^\alpha \ell_\alpha = -1$

so that $a = j/c$.

AIII.2. - Classical approximation [11]

Search for the main parts, with respect to c^{-2}, of the relativistic shock equations. We have first, according to (VII.3.2)

(AIII.2.1) $\quad [\, j \,] = [\, rw \,] = 0$

The invariance relation (VII.3.3) gives

$$(1 + i_1 c^{-2})\, h_1^\alpha \ell_\alpha = (1 + i_0 c^{-2})\, h_0^\alpha \ell_\alpha$$

It follows that, up to the terms in c^{-4}, one has

(*) $\quad h_1^\alpha \ell_\alpha = [\, 1 - (i_1 - i_0)\, c^{-2} \,]\, h_0^\alpha \ell_\alpha$

One obtains in the Newtonian approximation

(AIII.2.2) $$[\, h^{\alpha} \ell_{\alpha} \,] = 0$$

From (VII.3.6) it follows

$$[\, q + j^2 \tau \,] = 0$$

The Newtonian approximation of this relation gives

(AIII.2.3) $$[\, p + (1/2)\, \mu \,|h|^2 + r\, w^2 \,] = 0$$

According to (*), the invariance of (VII.3.4), let

$$II = (c^2 / j^2)\,(h^{\alpha} \ell_{\alpha})^2 - |h|^2 / r^2$$

gives, up to terms in c^{-2},

$$(c^2/j^2)(1 - 2\,(i_1 - i_0)\, c^{-2})(h_0^{\alpha} \ell_{\alpha})^2 - (\,|h_1^2|^2 / r_1^2\,) =$$
$$= c^2/j^2)\,(h_0^{\alpha} \ell_{\alpha})^2 - (\,|h_0|^2 / r_0^2\,)$$

It follows at the Newtonian approximation

$$[(2i\,(h^{\alpha} \ell_{\alpha})^2 / j^2) + (\,|h|^2 / r^2\,)] = 0$$

that is, according to (AIII.2.1), (AIII.2.2)

(AIII.2.4) $$[\, i + \frac{1}{2} \frac{|h|^2}{(h^{\alpha} \ell_{\alpha})^2}\, w^2 \,] = 0$$

One has $|h|^2 = |k|^2 + (h^\alpha \ell_\alpha)^2$ and (AIII.2.4) may be written

(AIII.2.4')
$$[\, i + \frac{1}{2} w^2 + \frac{1}{2} \frac{|k|^2}{(h^\alpha \ell_\alpha)^2} w^2 \,] = 0$$

Finally, consider the relation of invariance (VII.3.5). The quantity

$$c^{-2} \alpha = r^{-1} (1 + ic^{-2}) - \mu ((h^\alpha \ell_\alpha)^2 j^{-2} - |h|^2 r^{-2} c^{-2})$$

has for Newtonian approximation

$$r^{-1} - \mu (h^\alpha \ell_\alpha)^2 j^{-2}$$

On the other hand

$$\chi = |h|^2 - a^2 H = |h|^2 - (h^\alpha \ell_\alpha)^2 + a^2 |h|^2 r^{-2}$$

that is equivalent at the classical approximation

$$\chi = |k|^2 + j^2 |h|^2 r^{-2} c^{-2} \simeq |k|^2$$

Therefore (VII.3.5) gives

(AIII.2.5)
$$[\, |k| (r^{-1} - \mu (h^2 \ell_\alpha)^2 j^{-2}) \,] = 0$$

The five Eqs (AIII.2.1)(AIII.2.5) are the shock equations of the classical Magnetohydrodynamics for a frame connected with the shock front.

At the classical approximation, the relation (AII.2.4) corresponding to the special case may be written

(AIII.2.6)
$$r v^2 = \mu |h|^2$$

BIBLIOGRAPHY

[1] Y. Choquet-Bruhat, Astron. Acta t 6 , (1960) p 354-364.

[2] Y. Choquet-Bruhat, C.R. Acad. Sci. Paris t 258, (1964), p 3809-12.

[3] Y. Choquet-Bruhat, Comm. Math. Phys. t 12, (1969), p 16-35.

[4] F. Hoffman and E. Teller, Phys. Rev. t 80 , (1950), p 692-701.

[5] W. Israel, Proc. Roy. Soc. A t 259, (1960), p 129-135.

[6] J. Leray, Hyperbolic Diff. Equations Lectures Notes, Institute for Advanced Stud. Princeton (1953).

[7] A. Lichnerowicz, Propagateurs et commutateurs en relativité générale, Publ math. de l'I.H.E.S. n° 10, Presses universitaires de France, Paris (1961).

[8] A. Lichnerowicz, Relativistic hydrodynamics and magneto-hydrodynamics. Benjamin New-York (1967).

[9] A. Lichnerowicz, in Relativistic fluid dynamics (Cattaneo ed.) p 89-203. Cremonese Roma (1971).

[10] A. Lichnerowicz, C.R. Acad. Sci. Paris t 276, série I, (1973), p 1385-1390.

[11] A. Lichnerowicz, Journal of Math Phys t 17, (1976) p 2133-2142.

[12] A. Lichnerowicz, Relativity and Math Phys, in Relativity, Quanta and Cosmology (transl. from ital. ; F. de Finis ed.) t II, (1989), p 403-472.

[13] J.C. Lucquiaud, Thèse Paris (1966).

[14] R. Polovin, Shock waves in magnetohydrodynamics, Sov. Physics, t 3 , 1961), p 677-688.

[15] N. R. Sibgatullin, Oscillations and waves in strong gravitational and electromagnetic fields, Springer Berlin (1990).

[16] V. D. Zakharov, Gravitational waves in Einstein's theory. John Wiley New-York (1973).

NOTE

APPROACH TO A QUANTUM THEORY OF FIELDS FOR A CURVED SPACE-TIME

I - TENSOR PROPAGATORS

NI.1. - Orientations over a space-time.

Let (V_4, g) be a space-time of class C^∞. The results are independant of this precision, which facilitates the mathematical formulation.

The orthonormal frames $y = \{e_0, e_i\}(\alpha = 0, 1, 2, 3, 4 ; i = 1, 2, 3)$ of (V_4, g) are the elements of a principal fiber bundle $\hat{E}(V_4)$ with base V_4, admitting the complete Lorentz group $L(4)$, with four components, as structural group.

a) If $A = (A_\alpha^{\lambda'})$ is a matrix, element of $L(4)$, ***its total signature*** ε_A ***is*** ± 1 ***according as the sign of*** $\det A$ ***; its temporal signature*** ρ_A ***is*** ± 1 ***according as the sign of*** $A_0^{0'}$. These two signatures define the component of $L(4)$ to which A belongs. ***A total orientation*** ε of (V_4, g) is a geometric object defined over V_4, with respect to the elements of $\hat{E}(V_4)$ by a component $\varepsilon_y = \pm 1$, such that if $y = y'A$ in matrix terms, one has $\varepsilon_y = \varepsilon_{y'}$, ε_A. ***A temporel orientation*** ρ is defined over (V_4, g) by a component $\rho_q = \pm 1$ such that, if $y = y'A$, one has $\rho_y = \rho_{y'} \rho_A$. The existence of a total orientation for V_4 depends only on the structure of the manifold (an "orientable" manifold), the existence for a temporal orientation depends on the field of elementary cones. If there exists a total orientation ε of V_4, the manifold admits only two orientations ε and $-\varepsilon$; similarly for a temporal orientation [7].

We always suppose in the following that (V_4, g) ***has been provided with a total orientation*** ε ***and a temporal orientation*** ϱ. The temporal orientation defines over all the manifold, a continuous field of future half-cones C_x^+ $(x \in V_4)$ and a continuous field of past half-cones C_x^-.

b) A temporal path of (V_4, g) is a piecewise C^1-path such that the half-tangents at each point x are ***in or over*** C_x. Let K be a set of V_4; the future $\mathcal{E}^+(K)$ (resp. the past $\mathcal{E}^-(K)$) of K is the set of points of V_4 given by the temporal paths issuing from the points x of K in the future of x (resp. terminating at the points x of K in the past of x).

These definitions are valid in particular for $K = \{x'\}$. Set $\mathcal{E}(x') = \mathcal{E}^+(x') \cup \mathcal{E}^-(x')$. The boundary $\Gamma_{x'} = \partial\,\mathcal{E}(x')$, tangent at each point $x \neq x'$ to the cone C_x is called the characteristic conoid of x' ; it is generated by the isotropic geodesics issuing from x'. In a neighborhood of its vertex x', the conoid $\Gamma_{x'}$ is diffeomorphic to a neighborhood of an ordinary cone of second order. But the global phenomena are very different ; in particular, the isotropic geodesics issuing from x' may intersect themselves.

NI.2. - Global hyperbolicity.

a) The theory of linear systems of hyperbolic partial derivatives with variable coefficients leads to the introduction of the following hypothesis, which assures the existence of elementary solutions [4].

Global strict hyperbolicity of (V_4, g). ***Consider the space of paths of*** V_4 ***equipped with the compact-open topology. The set of temporal paths joining two arbitrary points of*** V_4 ***is assumed to be either empty or compact in the preceding topological space.***

According to J. Leray and Y. Choquet-Bruhat, this hypothesis entails in particular that ***the set of points of*** V_4, given by the temporal paths joining two arbitrary points ***is either empty or compact in*** V_4. One sees easily that throughout (V_4, g), there exists open submanifolds of V_4 containing an arbitrary point and which are globally hyperbolic strictly.

b) We may prove that global hyperbolicity may be expressed in a more geometrical way.

Hypothesis (H)

1) There is a real-valued C^∞-function φ over V_4 such that $\varphi = t = \text{const.}$ are space-like regular hypersurfaces Σ_t diffeomorphic to a manifold W_3 of dimension 3.

2) There is over (V_4, g) a smooth strictly time-like vector field Z with integral curves diffeomorphic to $\mathbb{R}$.

3) There is a diffeomorphism $f : V_4 \to W_3 \times \mathbb{R}$ which maps the leaves Σ_t onto $W_3 \times \{ t \}$ and the integral curves of Z onto $\{ y \} \times \mathbb{R}$ ($y \in W_3$)

(V_4, g) is thus globally strictly hyperbolic.

c) If we have ***global hyperbolicity,*** a set K of V_4 is said to be ***compact toward the past*** (Leray) if the intersection $\mathcal{E}^+(K) \cap \mathcal{E}^-(x)$ is either empty or compact for all points x of V_4 ; $\mathcal{E}^+(K)$ itself and any closed subset of $\mathcal{E}^+(K)$ are then also compact toward the past. The notion of compactness toward the future is defined similarly. Leray has established the following lemma : ***If*** K ***is compact toward the past and*** K' ***is compact*** $\mathcal{E}^+(K) \cap \mathcal{E}^-(K')$ ***is always compact*** (resp. If K is compact toward the future and K' is compact $\mathcal{E}^-(K) \cap \mathcal{E}^+(K')$ is always compact).

NI.3. - Bitensors and Dirac Bitensors [6]

a) In the theory of differential operators over V_4 , the notions of bitensors and bitensor distributions necessarily arise. If T_x is the vector space tangent at x to V_4 , we identify its dual T_x^* with T_x by means of the metric g . In this Note moreover we denote by δ the codifferentiation operator for (V_4, g) .

A (p, q)-bitensor at (x, x') is an element of the tensor product $(\otimes_p T_x) \otimes (\otimes_q T_{x'})$ A(p, q)-bitensor of V_4 is a field of (p, q)-bitensors over $V_4 \times V_4$.

A (p, q) ***bitensor distribution*** is a continuous linear form, with scalar values, over the smooth (p, q)-bitensors with compact support in

$V_4 \times V_4$ "Continuous" being intended in the sense of distributions. In particular, a continuous linear form, with p-tensor values, over the smooth q-tensors with compact support in V_4 defines a (p, q)-bitensor distribution called ***special.***

b) Consider for example the Dirac biscalar $\hat{\delta}(x, x')$ of V_4 defined as the special biscalar distribution given by

(NI.3.1) $$< \hat{\delta}(x, x') \; , \; f(x') > = f(x)$$

The support of $\hat{\delta}$ is the diagonal of $V_4 \times V_4$.

More generally, let $\tau(x, x')$ be a (1, 1)-bitensor, arbitrary but under the single condition that $\tau\ (x, x)$ defines the representative tensor of the identity operator at x. Taking the tensor product of τ with itself, one obtains (p, p)-bitensors $\otimes_p \tau$. ***The Dirac bitensors of*** (V_4, g) are the special (p, p)-bitensor distributions defined by

(NI.3.2) $$D^{(p)}(x, x') = (\otimes_p \tau)\, \hat{\delta}(x, x')$$

which are independant of the choice of τ ; $D^{(p)}$ may also be defined by the relation

(NI.3.3) $$< D^{(p)}(x, x') , U(x') > = U(x)$$

where U is a p-tensor.

Through antisymmetrization, we obtain a bi-p-form distribution $\hat{D}^{(p)}$ and by symmetrization of $D^{(2)}$ a symmetric bitensor distribution

$\tilde{D}^{(2)}$. If d_x is the operator of exterior differentiation with respect to x, δ_x the codifferentiation operator with respect to x, one proves immediately [6]

(NI.3.4) $$\delta_{x'} \tilde{D}^{p+1}(x, x') = d_x \tilde{D}^p(x, x')$$

If A is a vector field, we call $\mathscr{D} A$ the symmetric 2-tensor $\mathscr{L}(A)\, \mathbf{g}$ where $\mathscr{L}(A)$ is the Lie derivative with respect to A. In a manner analogous to (NI.3.4) one has

(NI.3.5) $$\delta_{x'} \tilde{D}^2(x, x') = \mathscr{D}_x D^{(1)}$$

NI.4. - Linear differential-tensor operators associated with g.

a) Introduce over $(V_4, \mathbf{g})$ the operator on p-tensors defined by

$$(\Delta T)_{\alpha_1 \dots \alpha_p} = - \nabla^\rho \nabla_\rho T_{\alpha_1 \dots \alpha_p}$$

where ∇ is the operator of covariant differentiation in the riemannian connection. We are moreover given a field F of linear operators F_x over p-tensors and a field B of linear mappings of (pH)-tensors over p-tensors ; B_x may be defined by a set b_x^ρ of operators on p-tensors. Denote by $B_x^{\rho *}$ and F_x^* the transposed operators with respect to $\mathbf{g}$. Consider the linear differential operator over p-tensors

(NI.4.1) $$LT = \Delta T + B^\rho \nabla_\rho T + FT$$

The formal adjoint operator L^* of L is given by

(NI.4.2) $$L^*U = \hat{\Delta} U - \nabla_\varrho (B^{\varrho *} U) + F^* U$$

In fact, if T is a p-tensor distribution, U a p-tensor such that $S(T) \cap S(U)$ is compact one has

(NI.4.3) $$< LT , U > = < T , L^*U >$$

In order that L be self-adjoint, it is necessary and sufficient that $B^{\varrho *} = - B^\varrho$ and $F^* = F - \nabla_\varrho B^\varrho$

b) Introduce over tensors a ***laplacian*** different than $\hat{\Delta}$ and more adapted to geometrical and physical applications. For antisymmetric tensors (or forms) G. de Rham has introduced the famous laplacian

(NI.4.4) $$\Delta T = (d\delta + \delta d)\, T$$

Δ commutes in this case with d and δ. One has explicitely in terms of the curvature and Ricci tensors

(NI.4.5) $$(\Delta T)_{\alpha_1 \dots \alpha_p} = - \nabla^\varrho \nabla_\varrho T_{\alpha_1 \dots \alpha_p} + \sum_\ell R^\varrho_{\alpha_\ell} T_{\alpha_1 \dots \varrho \dots \alpha_p}$$

$$- \sum_{k \neq \ell} R^{\varrho\ \sigma}_{\alpha_k\ \alpha_\ell} T_{\alpha_1 \dots \varrho \dots \sigma \dots \alpha_p}$$

I have proposed [6] to define the laplacian ΔT of an arbitrary p-tensor by formula (NI.4.5). The corresponding Klein-Gordon operator is $\Delta + \mu$, where $\mu = \text{const}$.

The operator Δ enjoys the following properties : it is self-adjoint, commutes with contraction and with the transposition operators on indices of T ; as a consequence, it preserves the possible symmetry or antisymmetry properties of T. If T is such that $\nabla T = 0$, $\Delta T = 0$ and if U is arbitrary

(NI.4.6) $$\Delta(T \otimes U) = T \otimes \Delta U$$

If the Ricci tensor is parallel, one has for any 2-tensor T and vector A

(NI.4.7) $$\delta \Delta T = \Delta \delta T \qquad \nabla \Delta A = \Delta \nabla A$$

NI.5. - Elementary Kernels and propagators.

We suppose in the following that (V_4 , g) is ***globally strictly hyperbolic and*** we consider the operator L defined by

$$L = \Delta T + B^{\varrho} \nabla_{\varrho} T + FT$$

a) We have the following classical theorem [4]

Theorem. - *Over the globally strictly hyperbolic manifold* (V_4 , g) , *there exist two elementary kernels* $E^{(p)\pm}(x, x')$ *of the operator* L *,that is two special* (p,p)*-bitensor distributions satisfying*

(NI.5.1) $$L^*_x E^{(p)\pm}(x, x') = D^{(p)}(x, x')$$

which, for each x', *have their support respectively in* $\mathcal{E}^+(x')$ *and* $\mathcal{E}^-(x')$. *These kernels are unique and also satisfy*

(NI.5.2) $$L_{x'} E^{(p)\pm}(x, x') = D^{(p)}(x, x')$$

b) The uniqueness of the elementary kernels is a particular case of a general uniqueness theorem [4] [6].

Uniqueness theorem. *Any tensor-distribution* T *, solution of the homogeneous equation* $LT = 0$ *with a support compact toward the future (resp. toward the past) is necessarily zero.*

In fact, consider a p-tensor distribution T , solution of the homogeneous equation

(NI.5.3) $$LT = 0$$

with a support $S(T)$ ***compact toward the future.***

To any smooth p-tensor V ***with a compact support*** associate a p-tensor defined by

(NI.5.4) $$W(x) = < E^{(p)+}(x, x') , V(x') >$$

W has its support $S(W)$ in the future $\mathcal{E}^+(S(V))$ of the support $S(V)$ and satisfies

$$L^*_x W(x) = < L^*_x E^{(p)+}(x, x') , V(x') > = < D^p(x, x') , V(x') > V(x)$$

It is a solution of

(NI.5.5) $$L^* W = V$$

$S(T)$ being compact toward the future, $\mathcal{E}^+ (S(V)) \cap \mathcal{E}^- (S(T))$ is a compact set K. Let α be a smooth function with a compact support and equal to 1 in a compact neighborhood of K ; (NI.5.3) implies

$$< LT , \alpha W > = 0$$

It follows

$$< T , L^*(\alpha W) > = < T , L^* W > = 0$$

that is, according to (NI.5.5)

$$< T , V > = 0$$

for any V with compact support. One has $T = 0$ and the uniqueness theorem is proved.

It follows in particular that the two elementary solutions $E^+_{x'}(x)$ and $E^-_{x'}(x)$ (with supports in $\mathcal{E}^+ (x')$ or $\mathcal{E}^- (x')$) and the two corresponding elementary kernels are unically determined by (NI.5.1) and the conditions of support.

c) Consider the equation

(NI.5.6) $$LW = V$$

where V is an arbitrary smooth p-tensor with compact support. One has

Proposition. *-A solution* W *of the equation* (NI.5.6) *with a support compact toward the future is necessarily the p-tensor*

(NI.5.7) $$W(x') = \langle E^{(p)+}(x, x') , V(x) \rangle$$

In fact, let W be a solution with support compact toward the future of (NI.5.6). L'intersection $\mathcal{E}^-(S(V)) \cap \mathcal{E}^+(x')$ is empty or is a compact K. Let α be a smooth function with compact support, equal to 1 in a compact neighborhood of K. For x' in the past of $S(V)$, one has

$$\langle E^{(p)+}(x, x'), V(x) \rangle = \langle E^{(p)+}(x, x'), L_x W(x) \rangle =$$

$$= \langle E^{(p)+}(x, x'), L_x (\alpha W)(x) \rangle$$

that is

$$\langle E^{(p)+}(x, x'), V(x) \rangle = \langle L_x^* \; E^{(p)+}(x, x'), \; \alpha(x) W(x) \rangle = W(x')$$

d) Denote by $E^{*(p)\pm}(x', x)$ the elementary kernels associated to the operator L^* adjoint to the operator L. One has

$$L_{x'} , E^{*(p)\pm}(x', x) = D^{(p)}(x', x)$$

Consider the p-tensor

(NI.5.8) $$W_1(x') = \langle E^{*(p)-}(x', x) , V(x) \rangle$$

where V is an arbitrary smooth tensor with compact support. The support of W_1 is in the past of $S(V)$ and one has

$$L_{x'} W_1(x') = \langle L_{x'} E^{*(p)-}(x', x), V(x) \rangle = \langle D^{(p)}(x', x), V(x) \rangle = V(x')$$

For each V with compact support, W_1 defined by (NI.5.8) coincides with W defined by (NI.5.7). It follows

(NI.5.9) $$E^{*(p)-}(x', x) = E^{(p)+}(x, x')$$

and similarly

(NI.5.10) $$E^{*(p)+}(x', x) = E^{(p)-}(x, x')$$

Therefore

(NI.5.11) $$L_{x'} E^{*(p)\pm}(x, x') = D^{(p)+}(x, x')$$

e) Consider the bitensor-distribution $E^{(p)}$ defined by

(NI.5.12) $$E^{(p)} = E^{(p)-} - E^{(p)+}$$

For each x', $E^{(p)}(x, x')$ has its support in $\mathscr{E}(x')$ and satisfies

(NI.5.13) $$L^*_x E^{(p)}(x, x') = 0$$

It also satisfies

(NI.5.14) $$L_{x'} E^{(p)}(x, x') = 0$$

$E^{(p)}$ ***is by definition the p-tensor propagator relative to the operator*** L. If $E^{*(p)}$ is the propagator relative to L^*, it follows from

(NI.5.9), (NI.5.10) that

$$E^{*(p)}(x', x) = -E^{(p)}(x, x')$$

If L *is self-adjoint,* one has

(NI.5.15) $E^{(p)\mp}(x, x') = E^{(p)\pm}(x, x') \qquad E^{(p)+}(x', x) = -E^{(p)}(x, x')$

and $E^{(p)}$ is antisymmetric with respect to x and x'.

NI.6. - Tensor propagators relative to the space-time of Minkowski.

Consider in particular the space-time of Minkowski and the de Rham-operator Δ relative to p-tensors ($p = 0, 1, 2, \ldots$)

a) In the scalar case, if $D^{(0)-}(x, x')$ and $D^{(0)+}(x, x')$ are the elementary kernels corresponding to Δ, $D^{(0)-}(x, x')$ satisfied for each x'

$$\Delta_x D^{(0)-}(x, x') = \hat{\delta}(x, x')$$

and for each x' has its support in the past of x'. Similarly

$$\Delta_x D^{(0)+}(x, x') = \hat{\delta}(x, x')$$

and $D^{(0)+}(x, x')$ has, for each x', its support in the future of x'.

In terms of Fourier transformation, Jordan and Pauli have introduced in this space-time a scalar-distribution

$$D^{(0)}(x) = D^{ret}(x) - D^{adv}(x)$$

where D^{ret} has its support in the future of the origin and D^{adv} in the past of the origin and where, with our definition of Δ , one has

$$\Delta_x D^{ret}(x) = -\hat{\delta}(x) \qquad\qquad \Delta_x D^{adv}(x) = -\hat{\delta}(x)$$

We set

$$D^{ret}(x, x') = D^{ret}(x-x') \qquad\qquad D^{adv}(x, x') = D^{adv}(x-x')$$

and

$$D^{(0)}(x, x') = D^{(0)}(x - x')$$

where $D^{ret}(x, x')$ has its support in the future of x' and $D^{adv}(x, x')$ in the past of x' . It follows from the uniqueness theorem that

$$D^{ret}(x, x') = -D^{(0)+}(x, x') \qquad D^{adv}(x, x') = -D^{(0)-}(x-x')$$

Therefore $D^{(0)}(x, x')$ ***is the scalar propagator associated, for the space-time of Minkowski, to the operator*** Δ . ***It is the Jordan-Pauli propagator.***

b) Study for this space-time, the tensor propagators relative to the operator Δ . For such a space, if U is an arbitrary tensor

$$\Delta U = -\nabla^{\rho} \nabla_{\rho} U$$

and, according to (NI.4.6), if T such that $\nabla T = 0$, one has

$$\Delta(T \otimes U) = T \otimes \Delta U$$

Suppose that in each point x (resp x') of V_4 corresponds a frame $\{ e_\alpha \}$ (resp. $\{ e_{\lambda'} \}$ and consider the bi-1-tensor of components $t_{\alpha\lambda'} = e_\alpha \cdot e_{\lambda'}$; t is the global bi-1-tensor of parallel transport which has a zero covariant derivative with respect to x and to x'.

Introduce the bi-p-tensors $(\otimes_p t)$. One has

$$\Delta_x [(\otimes_p t) D^{(0) \pm} (x, x')] = (\otimes_p t) \Delta_x D^{0 \pm} (x, x')$$

that is

$$\Delta_x [(\otimes_p t) D^{(0) \pm} (x, x')] = (\otimes_p t) \hat{\delta}(x, x') = D^{(p)} (x, x')$$

where t is a particular case of τ. It follows from the uniqueness theorem that the elementary kernels corresponding to Δ acting over the p-tensors are

$$E^{(p) \pm} (x, x') = (\otimes_p t) D^{(0) \pm} (x, x')$$

and for the tensor propagator

$$E^{(p)} (x, x') = (\otimes_p t) D^{(0)} (x, x')$$

We have the same formula for the propagators corresponding to the operator $(\Delta + \mu)$ $(\mu = \text{const.})$ if it is the case for $D^{(0)}$. Such results do not extend to the case of curved space-time.

NI.7. - Propagators relative to the operator $(\Delta + \mu)$

a) The self-adjoint Klein-Gordon operator $(\Delta + \mu)$ operates on antisymmetric p-tensors. One deduces from $E^{(p)\pm}$ by antisymmetrization two bi-p-form distributions $G^{(p)\pm}$ satisfying

(NI.7.1) $$(\Delta_x + \mu)\, G^{(p)\pm} = \hat{\delta}^{(p)}(x, x')$$

The difference $G^{(p)} = G^{(p)-} - G^{(p)+}$ is the antisymmetric propagator relative to $(\Delta + \mu)$; it satisfies

(NI.7.2) $$(\Delta_x + \mu)\, G^{(p)}(x, x') = 0$$

Δ_x and d_x commuting, one has

$$(\Delta_x + \mu)\, d_{x'}\, G^{(p+1)\pm}(x, x') = d_x\, \hat{\delta}^{(P)}(x, x')$$

But Δ_x and $\delta_{x'}$, commute and there results

$$(\Delta_x + \mu)\, \delta_{x'}\, G^{(p+1)\pm}(x, x') = \delta_{x'}\, \hat{\delta}^{(P+1)}(x, x')$$

From (NI.3.4) by subtraction

$$(\Delta_x + \mu)\,(\delta_{x'}\, G^{(p+1)\pm}(x, x') - d_x\, G^{(p)\pm}(x, x')) = 0$$

It then follows from the uniqueness theorem that

(NI.7.3) $$\delta_{x'}\, G^{(p+1)\pm}(x, x') = d_x\, G^{(p)\pm}(x, x')$$

and by subtraction [6]

(NI.7.4) $$\delta_x\,,\ G^{(p+1)} = d_x\, G^{(p)}$$

b) Consider the operator Δ over 2-tensors

$$(\Delta T)_{\alpha\beta} = -\nabla^{\varrho}\nabla_{\varrho} T_{\alpha\beta} + R_{\alpha\varrho} T^{\varrho}_{\beta} + R_{\varrho\beta} T^{\varrho}_{\alpha} - 2 R_{\alpha\varrho\,,\,\beta\sigma}\ T^{\varrho\sigma}$$

The operator $(\Delta + \mu)$ operates on symmetric 2-tensors. By symmetry one deduces from $E^{(2)\,\pm}$ two symmetric bitensor-distributions $K^{\pm}$ satisfying

(NI.7.5) $$(\Delta_x + \mu)\, K^{\pm}(x, x') = \tilde{\delta}^{(2)}(x, x')$$

The corresponding symmetric propagator is $K = K^{-} - K^{+}$.

Let Ω be the domain of a local chart $\{ x^{\alpha} \}$ (resp. Ω' the domain of a local chart $\{ x^{\lambda'} \}$). In these charts, for $x \in \Omega$ and $x' \in \Omega'$, NI.7.5) may be written explicitely

$$(\Delta_x + \mu)\, K^{\pm}_{\alpha\beta\,,\,\lambda'\mu'}(x\,,x') = (t_{\alpha\lambda'}\, t_{\beta\mu'} + t_{\alpha\mu'}\, t_{\beta\lambda'})\ \hat{\delta}(x, x')$$

Denote by $K^{\pm}_{\lambda'\mu'}$ and $K_{\lambda'\mu'}$ the contractions in x of $K^{\pm}_{\alpha\beta\,,\,\lambda'\mu'}$ and $K_{\alpha\beta\,,\,\lambda'\mu'}$

$$K^{\pm}_{\lambda'\mu'} = g^{\alpha\beta}\ K^{\pm}_{\alpha\beta\,,\,\lambda'\mu'} \qquad K_{\lambda'\mu'} = g^{\alpha\beta} K_{\alpha\beta\,,\,\lambda'\mu'}$$

They are bitensor-distributions, scalar in x and symmetric 2-tensor in x'. The operator Δ commuting with the contraction it follows

$$(\Delta_x + \mu)\ K^{\pm}_{\lambda'\mu'} = 2\, t_{\alpha'\mu'}\, t^{\beta}_{\mu'}\, \hat{\delta}$$

But

$$t_{\alpha\lambda'}\, t^{\beta}_{\mu'}\, \hat{\delta}\,(x, x') = g_{\lambda'\mu'}(x')\, \hat{\delta}\,(x, x')$$

One obtains

$$(\Delta_x + \mu)\ K^{\pm}_{\lambda'\mu'} = 2\, g_{\lambda'\mu'}\, \hat{\delta}$$

On the other hand, if $G^{(0)\pm}$ are the scalar kernels relative to the operator $(\Delta + \mu)$ one deduces from

$$(\Delta_x + \mu)\, G^{(0)\pm} = \hat{\delta}$$

that

$$(\Delta_x + \mu)\,(2\, g_{\lambda'\mu'}\, G^{(0)\pm}) = 2\, g_{\lambda\mu'}\, \hat{\delta}$$

By subtraction, one obtains, according to the uniqueness theorem :

$$K^{\pm}_{\lambda'\mu'}(x, x') = 2\, g_{\lambda'\mu'}(x')\ G^{(0)\pm}(x, x')$$

For the corresponding propagators, one has [6]

$$K_{\lambda'\mu'}(x, x') = 2\, g_{\lambda'\mu'}(x')\ G^{(0)}(x, x')$$

which may be translated by

(NI.7.6) $$\mathrm{Tr}_x\, K(x, x') = 2\, \mathbf{g}(x')\ G^{(0)}(x, x')$$

c) Suppose that the Ricci tensor of g is of zero covariant derivative. By reasoning just as above, one may deduce from (NI.3.5) and (NI.4.7) that [6]

(NI.7.7) $$\delta_{x'} K(x, x') = \mathcal{D}_x G^{(1)}(x, x')$$

Relations (NI.7.4), (NI.7.6), (NI.7.7) play an essential role for the construction of commutators for tensor fields.

II - APPLICATIONS TO QUANTIZATION PROBLEMS OVER A CURVED SPACE-TIME.

NII.1. - Commutator for vector Meson.

Let (V_4, g) be a given globally strictly hyperbolic space-time. We consider some examples of commutators.

a) Consider a vector field described by a 1-form α subject to the field equation (Yukawa equation for a field of spin 1, mass ε^2)

(NII.1.1) $$\delta\, d\alpha = \varepsilon^2\, \alpha \qquad \varepsilon^2 = \text{const.} > 0$$

ε^2 being $\neq 0$, (NII.1.1) implies that $\delta\, \alpha = 0$ and (NII.1.1) is equivalent to the system :

(NII.1.2) $$(\Delta - \varepsilon^2)\, \alpha = 0 \qquad (\Delta = d\delta + \delta d)$$

and

(NII.1.3) $$\delta\alpha = 0$$

Suppose now that α is a 1-form distribution with operator values, where these values are operators for a complex Hilbert space. We ssek to construct a commutator $[\alpha(x), \alpha(x')]$ $(x, x' \in V_4)$, that is a bi-1-form distribution X with scalar values, that for each x' has its support in $\mathcal{E}(x')$, is antisymmetric with respect to (x, x') and satisfied with respect to x the field system (NII.1.2), (NII.1.3).

In restricted relativity, one has classically in our notation

$$[\alpha(x)\ ,\alpha(x')] = (h/i)\ \{\ t\,D\,(x, x') - (1/\varepsilon^2)\ d_x\, d_{x'}\, D(x, x')\ \}\ \mathrm{Id}\,.$$

where Id. is the identity operator for the Hilbert space, t is the (1.1) bitensor defining absolute parallelism in the flat space-time, and D is the Jordan-Pauli scalar propagator relative to $(\Delta - \varepsilon^2)$

b) Consider, over our curved space-time, the bi 1-form distribution

$$X = G^{(1)} - 1/\varepsilon^2\ )\ d_x\, d_{x'}\ G^{(0)}$$

where $G^{(1)}$ and $G^{(0)}$ are propagators of orders 1 and 0 relative to $(\Delta - \varepsilon^2)$.

The support condition and the antisymmetry condition are satisfied. Moreover

$$(\Delta_x - \varepsilon^2)\ X = 0$$

and from (NI.7.4)

$$\delta_x X = \delta_x G^{(1)} - (1/\varepsilon^2)\ d_{x'}\, \Delta_x\, G^{(0)} = \delta_x G^{(1)} - d_{x'}\, G^{(0)} = 0$$

Thus [6]

(NII.1.4) $$[\alpha(x)\ ,\ \alpha(x')] = (\hbar/i)\ \{\ G^{(1)}(x, x') - (1/\varepsilon^2)\ d_x\, d_{x'}\, G^{(0)}(x, x')\ \}\ \mathrm{Id}\,.$$

gives a commutator compatible with (NII.1.1) and which coincides in restricted relativity with the usual commutator. If we set $F = d\alpha$, we obtain for F the commutator

$$[F(x) , F(x')] = (\hbar/i)\, d_x\, d_{x'}\, G^{(1)} (x, x')\, Id. \tag{NII.1.5}$$

NII.2. Commutator for a free electromagnetic field.

For a free magnetic field, one has for the potential 1-form (potential-vector) (NI.1.1) without the mass term ($\varepsilon^2 = 0$). Equation (NI.1.1) is then invariant under of electromagnetic gauge, and one may impose that α satisfies the Lorentz condition $\delta\alpha = 0$. Consider as equation for the potential

$$\Delta\alpha = 0 \tag{NII.2.1}$$

which implies

$$\Delta\delta\,\alpha = 0 \tag{NII.2.2}$$

and adopt for the commutator of the potential [6]

$$[\alpha(x) , \alpha(x')] = (\hbar/i)\, G^{(1)} (x , x')\, Id . \tag{NII.2.3}$$

where $G^{(1)}$ corresponds to the operator Δ . There results

$$[\alpha(x) , \delta_{x'}\, \alpha(x')] = (\hbar/i)\, \delta_{x'}\, G^{(1)} (x, x')\, Id. = (\hbar/i)\, d_x\, G^{(0)} (x, x')\, Id.$$

and also

(NII.2.4) $$[\delta_x \alpha(x) , \delta_{x'} \alpha(x')] = (\hbar/i) \; \Delta_x G^{(0)} (x, x') = 0$$

The commutator (NII.2.3) is compatible with (NII.2.1), but not with the Lorentz condition. The situation is strictly similar to that which appeared in restricted relativity. It suffices to submit the states to the supplementary condition $\delta\alpha |\Phi> = 0$, which is according to (NII.2.2), a condition on the initial data of a Cauchy problem.

Equation (NII.2.3) gives for the electromagnetic field $F = d\alpha$ a commutator that is formally (NII.1.5) and one sees that this commutator is strictly compatible with the Maxwell equations of a free field

$$dF = 0 \qquad\qquad \delta F = 0$$

In fact, from the differential relation (NI.7.4) one has

$$d_x d_{x'} G^{(1)} = d_x \delta_x G^{(2)} = \Delta_x G^{(2)} - \delta_x d_x G^{(2)} = - \delta_x d_x G^{(2)}$$

where $G^{(2)}$ is the antisymmetric propagator of order 2 relative to Δ From $d^2 = 0$, $\delta^2 = 0$, one may deduce that the commutator is indeed compatible with the Maxwell equations.

NII.3. - Commutator for a varying gravitational field with mass.

Consider over a given space-time (V_4 , g) an arbitrary variation ∂g of metric tensor described by a symmetric 2-tensor h ($h = \partial g$) ; to h we associate the 2-tensor $h' = \mu(h)$ with

$$h'_{\alpha\beta} = h_{\alpha\beta} - (1/2)\, h\, g_{\alpha\beta} \qquad\qquad (h = g^{\rho\sigma} h_{\rho\sigma})$$

a) The connection suffers a variation $Z^{\gamma}_{\alpha\beta} = \partial\Gamma^{\gamma}_{\alpha\beta}$. It may be shown by direct calculation that

$$Z^{\gamma}_{\alpha\beta} = (1/2)\,(\nabla_{\alpha}\, h^{\gamma}_{\beta} + \nabla_{\beta}\, h^{\gamma}_{\alpha} - \nabla^{\gamma}\, h_{\alpha\beta})$$

For the curvature tensor, one has in a local chart

$$R^{\alpha}_{\beta,\gamma\delta} = \partial_{\gamma}\,\Gamma^{\alpha}_{\beta\delta} - \partial_{\delta}\,\Gamma^{\alpha}_{\beta\gamma} + \Gamma^{\alpha}_{\varrho\gamma}\,\Gamma^{\varrho}_{\beta\delta} - \Gamma^{\alpha}_{\varrho\delta}\,\Gamma^{\varrho}_{\beta\gamma}$$

By variation one obtains

$$\partial R^{\alpha}_{\beta,\gamma\delta} = \partial_{\gamma}\, Z^{\alpha}_{\beta\delta} - \partial_{\delta}\, Z^{\alpha}_{\beta\gamma} + \Gamma^{\alpha}_{\varrho\gamma}\, Z^{\varrho}_{\beta\delta} + \Gamma^{\varrho}_{\beta\delta}\, Z^{\alpha}_{\varrho\gamma} - \Gamma^{\alpha}_{\varrho\delta}\, Z^{\varrho}_{\beta\gamma} - \Gamma^{\varrho}_{\beta\partial}\, Z^{\alpha}_{\beta\delta}$$

This formula may be written

$$\partial R^{\alpha}_{\beta,\gamma\delta} = \nabla_{\gamma}\, Z^{\alpha}_{\beta\delta} - \nabla_{\delta}\, Z^{\alpha}_{\beta\gamma}$$

By contraction, one obtains the variation of the Ricci tensor

$$\partial R_{\alpha\beta} = \nabla_{\varrho}\, Z^{\varrho}_{\alpha\beta} - \nabla_{\beta}\, Z^{\varrho}_{\alpha\varrho}$$

let

$$2\partial R_{\alpha\beta} = -\nabla^{\varrho}\,\nabla_{\varrho}\, h_{\alpha\beta} + \nabla_{\varrho}\,\nabla_{\alpha}\, h^{\varrho}_{\beta} + \nabla_{\varrho}\,\nabla_{\beta}\, h^{\varrho}_{\alpha} - \nabla_{\alpha}\,\nabla_{\beta}\, h$$

But, according to the Ricci identity

$$\nabla_{\varrho}\,\nabla_{\alpha}\, h^{\varrho}_{\beta} = \nabla_{\alpha}\,\nabla_{\varrho}\, h^{\varrho}_{\beta} + R_{\alpha\varrho}\, h^{\varrho}_{\beta} - R_{\alpha\varrho\,,\,\beta\sigma}\, h^{\varrho\sigma}$$

It results

$$2\partial R_{\alpha\beta} = -\nabla^{\rho}\nabla_{\rho} h_{\alpha\beta} + R_{\alpha\rho} h^{\rho}_{\beta} + R_{\beta\sigma} h^{\sigma}_{\alpha} - 2R_{\alpha\rho,\beta\sigma} h^{\rho\sigma} -$$

$$- \nabla_{\alpha}(\delta h')_{\beta} - \nabla_{\beta}(\delta h')_{\alpha}$$

that is [6]

$$2\partial R_{\alpha\beta} = \Delta h_{\alpha\beta} - (\mathcal{D}\,\delta h')_{\alpha\beta}$$

b) Suppose that the given metric tensor satisfies the Einstein equations

$$R_{\alpha\beta} = \lambda\, g_{\alpha\beta} \qquad (\lambda = \text{const.})$$

Set $k = \delta h'$. One has by product by ∇^{α}

$$2\nabla^{\alpha}\partial R_{\alpha\beta} = -(\Delta\delta h)_{\beta} - \nabla^{\alpha}(\nabla_{\alpha} k_{\beta} + \nabla_{\beta} k_{\alpha})$$

let, according to the definition of k,

$$-2\nabla^{\alpha}\partial R_{\alpha\beta} = (\Delta k)_{\beta} + \nabla^{\alpha}\nabla_{\alpha} k_{\beta} + \nabla^{\alpha}\nabla_{\beta} k_{\beta} - \frac{1}{2}\nabla_{\beta}\Delta h$$

But from the definition of the operator Δ, it follows

$$-\nabla^{\alpha}\nabla_{\alpha} k_{\beta} = (\Delta k)_{\beta} - \lambda k_{\beta}$$

and from the Ricci identity

$$\nabla^{\alpha}\nabla_{\beta} k_{\alpha} = \nabla_{\beta}\nabla_{\alpha} k^{\alpha} + \lambda k_{\beta}$$

One has

$$-2\nabla^{\alpha}\,\partial R_{\alpha\beta} = 2\lambda\, k_{\beta} - \nabla_{\beta}\left(\frac{1}{2}\,\Delta h - \nabla_{\alpha} k^{\alpha}\right)$$

But

$$g^{\rho\sigma}\,\partial R_{\rho\sigma} = \frac{1}{2}\,\Delta h - \nabla_{\alpha} k^{\alpha}$$

One obtains

$$\nabla^{\alpha}\left(\partial R_{\alpha\beta} - \frac{1}{2}\, g_{\alpha\beta}\, g^{\rho\sigma}\,\partial R_{\rho\sigma}\right) = -\lambda k_{\beta}$$

This relation may be also obtained by variation of the conservation identity of the Einstein tensor.

c) Consider over $(V_4\,,\,\mathbf{g})$ the field described by the symmetric 2-tensor **h** subject to the field equations

(NII.3.1) $$\partial R_{\alpha\beta} = \mu\, h_{\alpha\beta} \qquad (\mu = \text{const.})$$

The metric tensor **g** describes here macroscopic space-time and we consider **h** as defining ***the microscopic gravitational field.***

The theory of the field **h** may be developped in a manner strictement parallel to the theory of vector fields. The homologous equations will be systematically provided of the same numbers. From the identity of b , it follows that the field equations (NII.3.1) imply

$$(\mu - \lambda)\,\mathbf{k} = (\mu - \lambda)\,\delta\mathbf{h}' = 0$$

Suppose $\varepsilon^2 = 2(\mu-\lambda) \neq 0$ (presence of a mass term) ; (NII.3.1) is equivalent to the system equivalent to the system

(NII.3.2) $$(\Delta - 2\mu)\, h = 0$$

and

(NII.3.3) $$\delta h' = 0$$

Suppose now that h is a symmetric 2-tensor distribution with operator values. We seek to construct a commutator $[h(x)\,,\,h(x')]$ compatible with (NII.3.2) and (NII.3.3). The results of Pauli and Fierz in the flat theory of gravitation and the analogy with the vector field theory lead us to consider [6]

(NII.3.4) $$[h(x)\,,\,h(x')] = (\hbar/i)\,\{K(x,x') -$$

$$- g(x)\,g(x')\,G^{(0)}(x,x') - (1/\varepsilon^2)\mathcal{D}_x\,\mathcal{D}_{x'}\,G^1_{(x,x')}$$

where the propagators K , $G^{(0)}$, $G^{(1)}$ are relative to the operator $(\Delta-2\mu)$. The conditions of support and of antisymmetry with respect to (x, x') are satisfied. From the properties of Δ corresponding to an Einstein space (formula (NI.4.7)) our commutator is compatible with (NII.3.2). According to (NI.7.7) one has

$$\delta_{x'}\,K(x, x') = \mathcal{D}_x\,G^{(1)}(x, x')$$

On the other hand

$$\delta_{x'}\,(g\,(x')\,G^{(0)}(x, x') = -\,d_{x'}\,G^{(0)}(x, x')$$

Finally, one has

and according to the Ricci identity

$$-\nabla^{\lambda'} \nabla_{\mu'} G^{(1)}_{\beta\lambda'} = -\nabla_{\mu'} \nabla^{\lambda'} G^{(1)}_{\beta\lambda'} - \lambda\, G^{(1)}_{\beta\mu'}$$

let

$$-\nabla^{\lambda'} \nabla_{\mu'} G^{(1)}_{\beta\lambda'} = \nabla_{\mu'} (\delta_{x'} G^{(1)})_{\beta} - \lambda\, G^{(1)}_{\beta\mu'}$$

or

$$-\nabla^{\lambda'} \nabla_{\mu'} G^{(1)}_{\beta\lambda'} = \nabla_{\mu'} (\nabla_{\beta} G^{(0)}) - \lambda\, G^{(1)}_{\beta\mu'}$$

It follows

$$\delta_{x'} \mathscr{D}_{x'} G^{(1)} = (\Delta_{x'} - 2\lambda)\, G^{(1)} + d_x d_{x'} G^{(0)}$$

let, according to $\Delta_{x'} G^{(1)} = 2\mu\, G^{(1)}$ and to the definition of ε^2

$$\delta_{x'} \mathscr{D}_{x'} G^{(1)} = \varepsilon^2 G^{(1)} + d_x d_{x'} G^{(0)}$$

One obtains thus :

$$[h(x), \delta_{x'} h(x')] = (\hbar/i)\, \{\mathscr{D}_x G^{(1)} + g(x)\, d_{x'} G^{(0)} -$$

$$- (1/\varepsilon^2)(\varepsilon^2 \mathscr{D}_x G^{(1)} + \mathscr{D}_{x'} d_x d_{x'} G^{(0)})\}$$

let

$$[h(x), \delta_{x'} h(x')] = (\hbar/i)\{g(x)\, d_{x'} G^{(0)} - (1/\varepsilon^2)\, \mathscr{D}_x d_x d_{x'} G^{(0)}\}$$

According to (NI.7.6), one obtains

$$[h(x),\, \delta_{x'} h(x')] = 0$$

Our commutator is compatible with (NII.3.3). We have given a commutator compatible with the field equations that reduces to that of Pauli–Fierz in the flat case.

commutator compatible with the field equations that reduces to that of Pauli–Fierz in the flat case.

NII.4. - Commutator for a varying gravitional field without mass term.

We suppose in this Section that $\varepsilon^2 = 0$, $\mu = \lambda$ (no mass term). We will verify that Equation (NII.3.1) is then ***invariant under change of gravitational gauge*** $h \to h + \mathcal{D}A$ ***where*** A ***is an arbitrary vector.*** In fact if $h = \mathcal{D}A$ the Lie derivative of the Ricci tensor $R_{\alpha\beta} = \lambda g_{\alpha\beta}$ is

$$\mathcal{L}(A)\, R_{\alpha\beta} = \lambda\,(g_{\rho\beta}\,\nabla_\alpha A^\rho + g_{\alpha\rho}\,\nabla_\beta A^\rho\,) = \lambda\,(\nabla_\alpha A_\beta + \nabla_\beta A_\alpha\,)$$

let

$$\mathcal{L}(A)\;\; R_{\alpha\beta} = \lambda(\mathcal{D}A)_{\alpha\beta}$$

and the property is proved.

a) It is possible through such a change to impose on h the condition $\delta h' = 0$. Consider the variation $(h + \mathcal{D}A)$ of the metric tensor. One has

$$\delta(h + \mathcal{D}A)' = \delta h' - \nabla^\alpha\,(\nabla_\alpha A_\beta + \nabla_\beta A_\alpha\,) + \nabla_\beta\,\nabla_\alpha A^\alpha$$

It follows from the Ricci identity

$$\nabla^\alpha\,\nabla_\beta A_\alpha - \nabla_\beta\,\nabla_\alpha A^\alpha = \lambda A_\beta$$

and one has :

$$\delta(h + \mathcal{D}A)' = \delta h' - \nabla_\alpha\,\nabla^\alpha A_\beta - \lambda A_\beta$$

let

$$\delta(h + \mathcal{D}A)' = \delta h' + (\Delta\,-\,2\lambda)\,A$$

$$(\Delta - 2\lambda)\, A + \delta h' = 0$$

We may impose to **h** the condition $\delta h' = 0$ [6]

b) In absence of a mass term $(\varepsilon^2 = 0)$ we adopt as field equation for **h**

(NII.4.1) $$(\Delta - 2\lambda)\, h = 0$$

which implies

$$(\Delta - 2\lambda)\, \delta h = 0$$

and

$$(\Delta - 2\lambda)\, h = 0$$

From (NII.4.1) we deduce

(NII.4.2) $$(\Delta - 2\lambda)\, \delta h' = 0$$

The analogy with the electromagnetic case and the study of Section (NII.3) suggest to the commutator :

(NII.4.3) $$[h(x), h(x')] = (\hbar/i)\, \{ K(x, x') - g(x)\, g(x')\, G^{(0)}(x, x') \}$$

where the propagators correspond to the operator $(\Delta - 2\lambda)$. One has

$$[h(x), Tr_{x'}\, h(x')] = -(\hbar/i)\, 2g(x)\, G^{(0)}(x, x')$$

and

$$[h(x), \delta_{x'}\, h(x')] = (\hbar/i)\{\mathscr{D}_x\, G^{(1)}(x, x') + g(x)\, d_{x'}\, G^{(0)}(x, x')\}$$

$$[h(x),\delta_{x'} h(x')] = (\hbar/i)\{\mathscr{D}_x G^{(1)}(x,x') + g(x)\, d_{x'} G^{(0)}(x,x')\}$$

It follows

$$[h(x),\delta_{x'} h'(x')] = (\hbar/i)\mathscr{D}_x G^{(1)}(x,x')$$

and according to the expression of $\delta_x \mathscr{D}_x G^{(1)}$

$$[\delta_x h(x),\, \delta_{x'} h'(x')] = (\hbar/i)(\Delta - 2\lambda)\, G^{(1)} = 0$$

The commutator (NII.4.3) is compatible with (NII.4.1) but not with the pseudo Lorentz condition $\delta h' = 0$. The situation is simular to that of the electromagnetic case. It is again sufficient to subject the states Φ to the supplementary condition $\delta h' \,|\Phi\rangle = 0$ which arise from (NII.4.2) as a condition on the initial data of a Cauchy problem.

NII.5. - Creation. Annihilation operators [8].

Let (V_4, g) a given global strictly hyperbolic space-time in the sense of Section (NI.2). We use the notations of (NI.2)b. For simplicity, we consider here only the case of a real scalar field u satisfying the Klein-Gordon equation

(NII.5.1) $$(\Delta - \varepsilon^2)\, u = 0$$

where the operator $(\Delta - \varepsilon^2)$ admits the antisymmetrical propagator G. If u is operator valued, the corresponding commutator is

(NII.5.2) $$[u(x),\, u(x')] = (\hbar/i)\, G(x,x')\, \mathrm{Id}.$$

Cauchy hypersurface diffeomorphic to W (see (NI.2)), the solution of the Cauchy problem is given by

$$u(x) = \int_{\Sigma} \{ u(y)\, \partial_{\lambda} G(x, y) - G(x, y)\, \partial_{\lambda} u(y)\}\, d\sigma^{\lambda}(y) \qquad \text{(NII.5.3)}$$

where $d\sigma^{\lambda}$ is the vector element of area of the hypersurface Σ. One has $u(y) = \hat{u}(y)$ $\partial_0 u(y) = \hat{v}(y)$ (transverse derivative), where $\hat{u}$, $\hat{v}$ are Cauchy data given over Σ ; $(\hat{u}, \hat{v})$ plays the role of a point in a phase space.

Moreover, according to results of Y. Choquet-Bruhat [2], each solution of (NII.5.1) may be obtained by composition of the propagator G with a function f that may be chosen with a support compact toward the past and the future. For the solution generated by f, we write

$$u_f(x) = \int_{V_4} G(x, x')\, f(x')\, \eta(x') \qquad \text{(NII.5.4)}$$

where η is the riemannian volume element. Consider, on the space of the solutions, the natural 2-form

$$\Omega(u, u') = \int_{\Sigma} \{u(y)\, \partial_{\lambda} u'(y) - u'(y)\, \partial_{\lambda} (y) \}\, d\sigma^{\lambda}(y) \qquad \text{(NII.5.5)}$$

where the right member is independant of Σ. If (f, g) have compact supports in V_4, one has

$$\int_{V_4 \times V_4} G(x, x')\, f(x)\, g(x')\, \eta(x)\, \eta(x') = \Omega\, (u_f\,, u_g)$$

Moreno [11] has introduced suitable Sobolev spaces $K^{\ell} \simeq H^{\ell}(W) \times H^{\ell-1}(W)$ $(\ell \in R)$ and has written (NII.5.1) under an

Moreno [11] has introduced suitable Sobolev spaces $K^\ell \simeq H^\ell(W) \times H^{\ell-1}(W)$ ($\ell \in R$) and has written (NII.5.1) under an evolution form $\partial u/\partial t = T^{-1}(t)\, u$, where the spectrum of $T^{-1}(t)$ is imaginary, for $u \in K = K^{1/2}$. He has proved that Ω defines on the space K of solutions ***a strongly symplectic structure.***

b) Suppose that there exists on V_4 a real kernel $G_1\,(x, x')$, symmetric in (x, x'), solution of (NII.5.1) and satisfying, for each Cauchy hypersurface Σ, the convolution relation

$$\text{(NII.5.6)} \quad G(x, x') = \int_\Sigma \{G_1\,(x, y)\, \partial_\lambda\, G_1\,(x', y) - G_1(x', y)\, \partial_\lambda\, G_1(x, y)\,\}\, d\sigma^\lambda(y)$$

If $u \in K$, associate to u the solution u_1 (compare with (NII.5.3))

$$\text{(NII.5.7)} \quad u_1(x) = \int_\Sigma \{\, u(y)\, \partial_\lambda\, G_1\,(x, y) - G_1(x, y)\, \partial_\lambda\,(y)\,\}\, d\sigma^\lambda(y)$$

We define thus on K a linear automorphism $J : u \to u_1$ admitting the following properties

α) $J^2 = -\,\mathrm{Id}.$

β) $\Omega(Ju\,, Jv) = \Omega(u, v)$

Moreover, we impose

γ) $\Omega(u, Ju) > 0$ if $u \neq 0$

it is possible to see that (Ω, J) defines on K and infinite-dimensional Kähler stucture. If f , g have compact supports in V_4 , one has

$$< u_f , u_g > = \int_{V_4 \times V_4} G_1 (x, x') f(x) g(x') \eta(x) \eta(x')$$

Conservely it is easy to prove that if a Kähler stucture on (K, Ω) is generated by a distribution Kernel, this kernel has all the properties of G_1 .

The formula (NII.5.6) has been given in 1967 by A. Lichnerowicz (preprint). For a flat space-time the symmetric and antisymmetric kernels D_1 and D of Jordan-Pauli are effectively concerned by (NII.5.6). Moreover, in this case, the formula (NII.5.6) together with the condition of positivity, determines completely D_1 from D (Rideau).

c) We set

$$2\, G^{(+)} = G - i\, G_1 \qquad 2\, G^{(-)} = G + i\, G_1$$

$$2\, u^{(+)} = u - i\, u_1 \qquad 2\, u^{(-)} = u + i\, u_1$$

where $u^{(+)}$ (resp $u^{(-)}$) is a complex valued eigenfunctions of J corresponding to the eigenvalue i (resp. $-i$). The mainformula (NII.5.6) implies the three following results

1) (+) and (−) are projectors onto two supplementary eigenspaces.

2) $< u^{(+)} , u^{+} > = < u^{(-)} , u^{(-)} > = (1/2) < u , u >$

2) $< u^{(+)} , u^{+} > = < u^{(-)} , u^{(-)} > = (1/2) < u , u >$

3) If u is operator-valued, one has

$$[u^{(+)}(x) , u^{+}(x')] = [u^{(-)}(x) , u^{(-)}(x')] = 0$$

and

$$[u^{+}(x) , u^{(-)}(x')] = (\hbar/i)\ G^{(+)} (x, x')$$

$$[u^{-}(x) , u^{(+)}(x')] = (\hbar/i)\ G^{(-)} (x, x')$$

The projectors (+) and (−) correspond exactly to ***the partition of the field between positive and negative energies.*** A theory of the creation-annihilation operators can be deduced naturally from the formula (NII.5.6). All the usual propagators (in particular the Feynman propagator) may be deduced from the elementary kernels and from the propagator G_1.

Combet and Chevalier [1] [5] have proved by different methods that ***for a stationnary space-time*** there exists, G being known, one and only one G_1 giving a c.s.p.o. and compatible with the time invariance.

d) For an arbitrary globally hyperbolic time [11] Moreno has proved that there exists for (K, Ω) ***infinitesily many kernels*** G_1 ***generating c.s.p.o.*** and in particular J' s. ***Two such operators*** J and J' ***differ by a real linear symplectic transformation*** S ***of the flat symplectic space*** (K, Ω), that is $J' = S J S^{-1}$ where $\Omega (Su , Su') = \Omega(u, u')$ (see also [9])

NII.6. Dirac field. [7]

a) Let (V_4, g) be a space-time with two orientations (total orientation and temporal orientation). We denote by $L_0(4)$ the connected Lorentz group$\not\subset$; its universal covering that is a two-fold covering is the group denoted by Spin (4). Let E be the fiber bundle of the direct orthonormalized frames with structural group $L_0(4)$. The space-time admits ***a spin structure*** if E admits a two-fold covering $\mathcal{E}$ that is a principal fiber bundle of structural group Spin (4). The elements z of $\mathcal{E}$ are called spin frames. In an arbitrary dimension n, the existence of a spin structure for a hyperbolic manifold (V_n, g) is equivalent to the nullity of the second characteristic class of Stiefel-Whitney of the manifold (Haefliger). In dimension 4, the space-time (V_4, g) admits a spin structure if and only if the corresponding manifold is parallelisable.

Consider only orthonormalized frames for the space-time and introduce the 4×4 Dirac matrices satisfying

$$\text{(NII.6.1)} \qquad \gamma_\alpha \gamma_\beta = \gamma_\beta \gamma_\alpha = -2\, g_{\alpha\beta}\, e \qquad (\alpha, \beta = 0, 1, 2, 3)$$

where $\gamma_\alpha = (\gamma^a_{\alpha b})$ (a, b = 1, 2, 3, 4). The group Spin (4) may be represented by a group of 4×4 matrices Λ satisfying

$$\text{(NII.6.2)} \qquad \Lambda \gamma_\alpha \Lambda^{-1} = A^{\beta'}_\alpha \gamma_{\beta'}$$

where $A = (A^{\beta'}_\alpha)$ is the projection of $\Lambda \in \text{Spin}(4)$ on $L_0(4)$.

A 1-spinor contravariant ψ is represented with respect to the spin frame z in $x \in V_4$ by (ψ^a) and by action of Spin(4) over $\mathcal{E}$, onehas in

A 1-spinor contravariant ψ is represented with respect to the spin frame z in $x \in V_4$ by (ψ^a) and by action of Spin(4) over $\mathcal{E}$, onehas in terms of matrices

$$\psi(z\Lambda^{-1}) = \Lambda\, \psi(z) \qquad \Lambda \in \text{Spin}(4)$$

Similarly a covariant 1-spinor φ is represented with respect to the spin frame by (φ_a) and one has

$$\psi(z\Lambda^{-1}) = \varphi(z)\, \Lambda^{-1}$$

Denote by S_x the space of the contravariant 1-spinors at $x \in V_4$, by $\overline{S}_x$ the space of the covariant 1-spinors at x, by T^c_x the complex tangent space to V_4 at x. By tensor products of these spaces one obtains the notion of tensor-spinor. In particular, according to (NII.6.2), the elements of the Dirac matrices are the components of a canonical 1-form- (1, 1) spinor γ, the main 1-form-spinor of (V_4, g). The product $\mathcal{A}\mathcal{C}$ of the charge conjugation $\mathcal{C}$ by the Dirac adjunction (depending on the temporal orientation) defines a spin 2-form Γ.

b) We suppose (V_4, g) referred to local sections of E (orthonormalized frames). We denote by $\omega_U = (\omega^\alpha_\beta)$ or $(\omega^{\alpha\beta})$ the 1-form of riemannian connexion for a domain U. According to the canonical isomorphism between the Lie algebras of $L_0(4)$ and Spin (4), the riemannian connection of (V_4, g) defines a natural spin-connection for the principal fiber bundle $\mathcal{E}$. This connection is given over U by the 1-form

$$\sigma = -(1/4)\, \omega^{\alpha\beta}\, \gamma_\alpha\, \gamma_\beta$$

contravariant 1-spinor field, the Dirac operator is given by

(NII.6.3) $$P\psi = \gamma^{\alpha} \nabla_{\alpha} \psi$$

The operator $\Delta = P^2$, laplacian of the spinors may be written

(NII.6.4) $$\Delta = -\nabla^{\rho} \nabla_{\rho} + (1/4) R$$

The considerations given for the tensors may be extended to the spinors. It is the case for Dirac bispinors $\Sigma^{1/2}(x, x')$ 1-spinor contravariant at x and 1-spinor covariant at x' and for the elementary kernels of the operator $(\Delta - \varepsilon^2)$ (ε = const) on the contravariant 1-spinors : there exist two elementary kernels $G^{(1/2)\pm}$ of the operator $(\Delta - \varepsilon^2)$, that are two bi-1-spinors-distributions contravariants at x and covariants at x' satisfying

(NII.6.5) $$(\Delta_x - \varepsilon^2)\, G^{(1/2)\pm}(x, x') = \Sigma^{(1/2)}(x, x')$$

which have, for each x', their supports respectively in $\mathcal{E}^+(x')$ and $\mathcal{E}^-(x')$.

If ψ is contravariant , φ covariant, denote $\mathcal{A}\psi$ by $\overline{\psi}$ (that is covariant) and $\mathcal{A}\varphi$ by $\overline{\varphi}$ (that is contravariant)

If $G^{(1/2)} = G^{(1/2)+} - G^{(1/2)-}$, one has by difference of (NII.6.5)

(NII.6.6) $$(\Delta_x - \varepsilon^2)\, G^{(1/2)}(x, x') = 0$$

and $G^{(1/2)}(x, x')$ is the propagator of $(\Delta - \varepsilon^2)$, with a support for each

and $G^{(1/2)}(x, x')$ is the propagator of $(\Delta - \varepsilon^2)$, with a support for each x' in $\mathcal{E}^+(x') \cup \mathcal{E}^-(x')$.

It is easy to prove that

(NII.6.7) $$\overline{G}^{(12)}(x', x) = -G^{(1/2)}(x, x')$$

that corresponds, in the spinor case, to the "antisymmetry" of the propagator.

b) Introduce on the contravariant spinors the operators

(NII.6.8) $$(P - \varepsilon)\psi \qquad (P + \varepsilon)\psi \qquad \varepsilon = \text{const} > 0$$

One has

(NII.6.9) $$(P - \varepsilon)(P + \varepsilon)\psi = (\Delta - \varepsilon^2)\psi$$

Similarly, to the covariant spinors correspond the operators

$$(\overline{P} - \varepsilon)\varphi \qquad (\overline{P} + \varepsilon)\varphi$$

where $\overline{P} = -\nabla_\alpha \varphi \gamma^\alpha$; we have also

(NII.6.10) $$\overline{\Delta}\varphi = -\nabla^\varrho \nabla_\varrho \varphi + (1/4)R\,\varphi$$

A free Dirac field is a field described by a contravariant spinor satisfying the field equation

Its Dirac adjoint $\overline{\psi}$ satisfies

(NII.6.11) $$(\overline{P}-\varepsilon)\,\overline{\psi} = 0$$

Consider the elementary ekrnels $G^{(1/2)\pm}(x, x')$ of $(\Delta-\varepsilon^2)$ and introduce the bi-1-spinor contravariant in x, covariant in x', given by

$$S^{(1/2)\pm}(x, x') = (P_x + \varepsilon)\, G^{(1/2)\pm}(x, x')$$

One has also :

$$S^{(1/2)\pm}(x, x') = (\overline{P}_{x'} + \varepsilon)\, G^{(1/2)\pm}(x, x')$$

One deduces from (NII.6.9)

$$(P_x - \varepsilon)\; S^{(1/2)\pm}(x, x') = \Sigma^{(1/2)}(x, x')$$

and the $S^{(1/2)\pm}(x, x')$ are the elementary kernels corresponding to the operators $(P-\varepsilon)$ and $(\overline{P}-\varepsilon)$. It follows by difference that [7]

$$S^{(1/2)} = (P_x + \varepsilon)\, G^{(1/2)}(x, x') = (\overline{P}_{x'} + \varepsilon)\, G^{(1/2)}(x, x')$$

satisfies

(NII.6.12) $$(P'_x - \varepsilon)\, S^{(1/2)}(x, x') = 0 \qquad (\overline{P}_{x'} - \varepsilon)\, S^{(1/2)}(x, x') = 0$$

The study of the free Dirac field in restricted relativity lead us to adopt as anti commutator for our field

adopt as anti commutator for our field

$$[\ \psi(x)\ ,\ \overline{\psi}(x')\ ]_{+} = (\hbar/i)\ S^{(1/2)}(x, x')\ e \qquad \text{(NII.6.13)}$$

In fact (NII.6.13) satisfies the usual condition of support, is strictly compatible with the Dirac equations (NII.6.10), (NII.6.11) and is invariant by charge conjugation.

Considerations similar to the considerations concerning a real scalar field may be developped to obtain a theory of creation-annihalation operators.

REFERENCES FOR THE NOTE.

[1] M. Chevalier, J. Math. Pures et Appl. 53, (1974), p 223.

[2] Y. Choquet-Bruhat, Propagateurs et solutions d'équations homogènes hyperboliques, C.R. Acad. Sci Paris 251, (1960), p 29-32.

[3] Y. Choquet-Bruhat, Solutions élémentaires du second ordre in Coll. int. CNRS de Nancy, 1956. Publ. CNRS, 1957.

[4] Y. Choquet-Bruhat, in Batelle Rencontres 1967. Lect. in Math and Phys. Benjamin New-York 1969.

[5] E. Combet, Séminaire Physique Mathématique Colège de France, 1965.

[6] A. Lichnerowicz, Propagateurs et commutateurs en relativité générale. Publ. Math. de l'Institut des Hautes Etudes Scientifiques n° 10, Presses Universitaires de France, Paris 1961.

[7] A. Lichnerowicz, Champs spinoriels et propagateurs en relativité générale, Bull. Soc. Math. France 92, (1964), p 128-227.

[8] A. Lichnerowicz, Proc. of the Steklov Inst. of Math. Russ. Ed 135 (1975), English Ed 1, 1978.

[9] B. Kay, Comm. Math. Phys. 62, (1978), p 55-70.

[10] C. Moreno, J. Math. Phys. 18, (1977), p 2153-2162 ; 19, (1978), p.92-99.

[11] C. Moreno, On the spaces of positive and negative frequency solutions of the Klein-Gordon equation in curved Space-times, Reports on Math. Phys. 17 (1980), p 333-358, man ree in 1976.

[12] J.E. Segal, Representations of the canonical commutation relations, Cargese 1967.

Mathematical Physics Studies

Publications

1. F.A.E. Pirani, D.C. Robinson and W.F. Shadwick: *Local Jet Bundle Formulation of Bäcklund Transformations.* 1979 ISBN 90-277-1036-8
2. W.O. Amrein: *Non-Relativistic Quantum Dynamics.* 1981 ISBN 90-277-1324-3
3. M. Cahen, M. de Wilde, L. Lemaire and L. Vanhecke (eds.): *Differential Geometry and Mathematical Physics.* Lectures given at the Meetings of the Belgian Contact Group on Differential Geometry held at Liège, May 2–3, 1980 and at Leuven, February 6–8, 1981. 1983 ISBN 90-277-1508-4 (pb)
4. A.O. Barut (ed.): *Quantum Theory, Groups, Fields and Particles.* 1983 ISBN 90-277-1552-1
5. G. Lindblad: *Non-Equilibrium Entropy and Irreversibility.* 1983 ISBN 90-277-1640-4
6. S. Sternberg (ed.): *Differential Geometric Methods in Mathematical Physics.* 1984 ISBN 90-277-1781-8
7. J.P. Jurzak: *Unbounded Non-Commutative Integration.* 1985 ISBN 90-277-1815-6
8. C. Fronsdal (ed.): *Essays on Supersymmetry.* 1986 ISBN 90-277-2207-2
9. V.N. Popov and V.S. Yarunin: *Collective Effects in Quantum Statistics of Radiation and Matter.* 1988 ISBN 90-277-2735-X
10. M. Cahen and M. Flato (eds.): *Quantum Theories and Geometry.* 1988 ISBN 90-277-2803-8
11. Bernard Prum and Jean Claude Fort: *Processes on a Lattice and Gibbs Measures.* 1991 ISBN 0-7923-1069-1
12. A. Boutet de Monvel, Petre Dita, Gheorghe Nenciu and Radu Purice (eds.): *Recent Developments in Quantum Mechanics.* 1991 ISBN 0-7923-1148-5
13. R. Gielerak, J. Lukierski and Z. Popowicz (eds.): *Groups and Related Topics.* Proceedings of the First Max Born Symposium. 1992 ISBN 0-7923-1924-9
14. A. Lichnerowicz, Magnetohydrodynamics: Waves and Shock Waves in Curved Space-Time. 1994 ISBN 0-7923-2805-1
15. M. Flato, R. Kerner and A. Lichnerowicz (eds.): *Physics on Manifolds.* 1993 ISBN 0-7923-2500-1
16. H. Araki, K.R. Ito, A. Kishimoto and I. Ojima (eds.): *Quantum and Non-Commutative Analysis.* Past, Present and Future Perspectives. 1993 ISBN 0-7923-2532-X

Kluwer Academic Publishers – Dordrecht / Boston / London

MIX
Papier aus verantwortungsvollen Quellen
Paper from responsible sources
FSC® C105338

If you have any concerns about our products,
you can contact us on
ProductSafety@springernature.com

In case Publisher is established outside the EU,
the EU authorized representative is:
Springer Nature Customer Service Center GmbH
Europaplatz 3, 69115 Heidelberg, Germany

Printed by Libri Plureos GmbH
in Hamburg, Germany